2/07

The Creative Enterprise

The Creative Enterprise
Managing Innovative Organizations and People

STRATEGY
VOLUME 1

Edited by
Tony Davila
Marc J. Epstein
and
Robert Shelton

Praeger Perspectives

Westport, Connecticut
London

Library of Congress Cataloging-in-Publication Data

The creative enterprise : managing innovative organizations and people / edited by
Tony Davila, Marc J. Epstein, and Robert Shelton.
    p. cm.
    Includes bibliographical references and index.
    ISBN 0-275-98685-3 (set : alk. paper) — ISBN 0-275-98686-1 (vol. 1 : alk. paper) —
ISBN 0-275-98687-X (vol. 2 : alk. paper) — ISBN 0-275-98688-8 (vol. 3 : alk. paper)
    1. Organizational change—Management. 2. Technological innovations—Management.
3. Creative ability in business—Management. 4. Industrial management. I. Davila, Tony.
II. Epstein, Marc J. III. Shelton, Robert D.
HD58.8.C727   2007
658.4′063—dc22   2006030628

British Library Cataloguing in Publication Data is available.

Library of Congress Catalog Card Number: 2006030628
ISBN: 0-275-98685-3 (set)
        0-275-98686-1 (vol. 1)
        0-275-98687-X (vol. 2)
        0-275-98688-8 (vol. 3)

First published in 2007

Praeger Publishers, 88 Post Road West, Westport, CT 06881
An imprint of Greenwood Publishing Group, Inc.
www.praeger.com

Printed in the United States of America

The paper used in this book complies with the
Permanent Paper Standard issued by the National
Information Standards Organization (Z39.48-1984).

10 9 8 7 6 5 4 3 2 1

Contents

Introduction

Business forces are eroding static competitive advantages faster than ever. And this is not only true for technology markets, where the pace has just accelerated. It is also true in industries that were considered "mature." Mittal, the steel company, is revolutionizing its industry. And its advantage does not come from amazing new technology, but from a relentless focus on doing business differently. Procter & Gamble has made explicit its compromise with innovation as the only way to remain profitable. This compromise with new technologies and business practices has already meant the resignation of one CEO, but not because he was too slow. Rather, he went too fast.

Innovation has emerged as the only way to sustain competitive advantage over time. Success is not to be found in a technology, in a market position, or in a business model; success resides in an organization's ability to innovate and be ahead of its competitors. This three-volume set is designed to provide the reader with the most up-to-date knowledge on how to be innovative. It addresses this issue from the various perspectives that are needed to have a well-rounded understanding of how to drive innovation in an organization.

The first volume takes a strategy perspective to answer the question of how to design an organization to be competitive in its market space. Innovation is not something that a manager can turn on only when needed. It is not a faucet that can be shut off when we don't need innovation and turned on when we do. Innovation is both a state of mind and a way of life. The first volume explores this idea from different perspectives on strategy.

The second volume looks at innovation from the perspective of the individual. It addresses the question of how to design organizations to enhance creativity. This volume focuses on drivers of creativity at the individual and team levels. Then it moves up a level of analysis and looks at organizational forces that shape this creativity—culture and rewards.

The third volume is about execution. It answers the question of how to get innovation done. The focus of this volume is how to design the management infrastructure to encourage innovation. Using a car race metaphor, the second volume is about the driver; this third one is about the car. The chapters address different tools to enhance innovation, from organizational structures to processes and measures.

The three volumes combine the perspective of large companies and small start-ups. Innovation is not the exclusive territory of one set of organizations. It happens in large companies as well as young ones; it happens in for-profit companies as well as not-for-profit organizations—under the umbrella of social innovation. The three volumes combine these various sources of innovation.

VOLUME 1: DEFINING INNOVATION STRATEGIES

Innovation starts at the top of an organization. It is top management's compromise with innovation that drives it. The first chapter in this volume shows how companies following an innovation strategy have outperformed more conservative ones. The chapter presents evidence from research studies and company stories to illustrate the importance of innovation to success.

Top management's commitment to innovation shows up in many different aspects. The chapters in Volume 1 address the aspects that make an innovative enterprise. The first aspect is the design of the *organization's interfaces with the environment*.

A key finding in both academic research and managerial practice is that innovation is not an individual activity—the popular image of the lone genius coming up with the most amazing ideas in a garage is a gross and dangerous simplification. Innovation—moving ideas into value—is a team effort. Ideas emerge and improve through exposure. The not-invented-here syndrome, where anything from the outside of a limited group is seen as inferior, is one of the most dangerous organizational pathologies. Top management is in charge of encouraging the interaction among people from different departments, bringing in people with different backgrounds, and ensuring the fluidity of ideas from outside the organization. One of the chapters in the first volume provides an interesting story on how innovation has happened in history. After reading this chapter, the reader will see innovation in a different light and understand how personalities, groups, and the environment interacted to deliver some of the most important innovations of the twentieth century.

In this first volume, three chapters cover the importance of the environment to innovation. One of them examines how Silicon Valley is redefining itself to maintain its undisputed leadership as the world's innovation hub. The chapter delineates the dynamism linked to people with different trainings interacting to create. Innovation in Silicon Valley is a team sport, with constant fluidity of ideas and backgrounds. Isolated companies have no room in the Valley. The second chapter takes the perspective of a university—one of the main sources of technological innovation—and its experiences with the corporate world. The chapter provides an interesting discussion on how technology-transfer offices work and the challenges they are facing to become more effective in moving technology breakthroughs to society. The third chapter also looks at the interface of the university and industry—a key link in leveraging the knowledge generated in universities. It presents a study on Engineering Research Centers: an organizational form that the National Science Foundation developed to improve technology commercialization at universities. The chapter details what makes some of these centers more successful.

Another aspect of innovation management that top management is in charge of is *defining the organization's innovation strategy*. Innovation is often confused with freedom. Providing direction and guidelines, setting criteria, and telling people what not to do are seen as ingredients to kill innovation. Much like the lone innovator, the need for unrestricted freedom to innovate is a myth. If top management wants innovation, it needs to set the strategy—decide what not to do and where the company needs to go. The CEO of Logitech—the leading company in computer devices such as mice and keyboards—provides a good example of giving directions and defining what is not within the company's strategy. He describes his company's strategy as "dominating the last inch," the inch that puts a person in contact with technology. So the company is not interested in technology products or in software products; it is interested in technology and software that facilitate the person-machine interaction. Logitech's CEO believes that this is a large enough space.

Three chapters in this first volume address the strategic dimension of innovation management. One of them provides a framework on how to think about innovation strategies. It describes the various levers that top management use to shape strategy. A second one addresses the important distinction between incremental and radical innovations. Incentives, risk aversion, and organizational antibodies lead to an emphasis on incremental innovation—more visible and profitable over the short term, but with the risk of jeopardizing the long term. The need for radical innovation and how to manage radical innovations are issues addressed in these chapters. While too much incremental innovation is dangerous, the opposite is also true. The right amount of innovation and the right mix are unique to every organization and where they are in their development. The third chapter addresses different ways in which management knowledge has thought about innovation strategy—how it has evolved from the idea of strategy as a plan designed by top management and implemented by the organization to the idea of innovation happening throughout the organization with top management being in charge of guiding and structuring these efforts. The evolution of the concept of strategy has led to changes in the way strategy implementation is executed.

Two chapters in the first volume address two important topics related to innovation. The first one presents the idea of social innovation—innovation in social settings, often through non-for-profit organizations. The advances in this topic of innovation have been amazing over the last few years. The world of social organizations has seen a management revolution as donors with deep managerial experience have adopted best practices in commercial companies as well as social organizations. In the academic world, a topic that was hardly taught has become one of the most popular courses in business schools. Stanford Graduate School of Business has launched a Social Innovation Center that publishes a magazine focused on the topic; it also offers several electives to MBAs and executive programs for non-profit organizations' leaders. The chapter addresses this important topic and examines how to adapt what we know about innovation in for-profit companies to social innovation.

The second important topic covered in this initial volume is innovation in start-up firms. The paradox here is that when talking about innovation, some people only think about how to make large firms more innovative, while others believe that only start-ups are innovative. The truth is that innovation happens in both types of organizations. This chapter discusses the evolution of start-up firms. A key transition point for these companies happens when their size is such that professional management tools are needed to implement strategy. The company is not a group of friends who can be managed as a group; it becomes an organization. Entrepreneurs often have a difficult time making this transition, and often they are replaced to bring in a manager. This chapter focuses on this transition point and how successful start-ups make this transition.

VOLUME 2: IMPROVING INNOVATION THROUGH PEOPLE AND CULTURE

The innovation lever addressed in the second volume is the internal environment. The amount of innovation within an organization depends, to a large extent, on top management's ability *to create the right culture and the right setting for people's creativity to thrive.* The volume starts by looking at what makes people creative. The first chapters describe in detail what we know about creativity and how to fully use the creative potential of people.

Creativity at the individual level has been the focus of much recent research. The conclusions from this research provide a complex picture, even more when creativity happens in an organization with different forces acting upon it. The need to transform ideas into useful solutions creates additional tensions in organizations. These tensions require balancing acts and a commitment from top management to let people run with ideas with a fuzzy future. The more novel an idea, the harder it is to visualize where it leads and the more fragile it is. Ideas need a runway to develop and an encouraging environment without premature judgments or negative feelings. They need experiments and prototypes to manage uncertainty. The planning is about how to resolve uncertainty, rather than visualizing the future, which is the practice with which we commonly associate planning.

Creativity is not about creating a perfect state; it is about balancing different forces over time. Positive and negative affective states, extrinsic and intrinsic motivation, autonomy, and guidance are required.

A recurring theme is the importance of the environment beyond individual creativity traits. People who could be considered less creative will outperform creative individuals if they have a supportive environment that the latter do not have. The characteristics of this environment range from leadership to co-workers. A person will be more creative when her supervisor does not micromanage and leaves space for ideas to emerge and mature, when the supervisor provides inspiration and stimulates innovation through, for instance, goals that demand creativity, when this person is fair and supportive in her evaluations. Similarly, co-workers who are creative and value creativity put together an environment where people thrive.

Contrary to common wisdom, creativity requires discipline—not the military discipline that eliminates it, but the discipline of working on it. Creativity does not just happen; people and organizations need to want it to happen. A key component of creativity is openness to experience, interacting with the outside world, with people with different experiences and points of view. Some people have a natural tendency to interact with "weird" people; but most of us prefer the safety of what we know. Discipline is required to overcome these creative blocks. Another component of creativity is to consciously think about these experiences and make the effort to translate them into ideas. Again, our natural tendency is to let these experiences go by, without considering how they can enrich the way we live and work.

Another important ingredient of creativity is self-confidence. Often, we are not creative because we do not believe we can be so. We don't even try to come up with new ways to look at the world. Several personal attitudes are blocks to new ideas, from having doubts about trying to think differently to fear of failing. Failure and creativity come together; actually, failure happens more often than success when risks are taken. In the same way that organizations that penalize failure will kill innovation, fear of failure kills the risk-taking attitude required for creativity.

The initial stage in formally tackling creativity is idea generation. At this stage, there should be no limits to what comes into the process. To do this, people involved have to forget about their self-image and their fear of saying something wrong—what other people are going to think. The richer this initial step, the better the raw material available. It is only as this process progresses that this raw material is processed into feasible ideas.

From individual creativity, the volume progresses into the topic of organizational culture and the social context of innovation. Certain organizations are more innovative than others. Strategy, as described in Volume 1, accounts for part of it. The informal norms and codes of conduct, what is broadly understood as culture, account for another important part. Finally, management infrastructure—the focus of Volume 3—accounts for the rest.

Culture has always fascinated managers and researchers in organizations. A culture that supports innovation is a culture that encourages people to interact with their networks to identify opportunities. It also provides resources and recognition to people who take risks exploring new ideas. It is a culture that recognizes effort and failure—a key ingredient of innovation. More importantly, innovative cultures tend to be strong cultures—cultures that reinforce and live very clear values and objectives. Clear values shift the attention from short-term financial performance to consistency with these values over time.

An innovative culture supports autonomy—where people can experiment—and risk taking. It has bias for action; rather than waiting for things to happen, an innovative culture will support people experimenting and prototyping their ideas. It has a winning mentality, with the objective of leading the market and achieving goals that seemed to be unreachable. It values openness to the world to enrich the idea generation process and values teamwork where ideas are bounced and refined. It is a culture that does not kill dissenting views but rather encourages the different points of view.

But culture goes beyond the organization to the level of nations. Certain nations are more innovative than others. The economic well-being, an appreciation for scientific work, a robust educational system, and the size of the nation all affect the level of innovativeness of a nation.

Finally, the second volume addresses the process of innovation—how to design such a process to enhance individual creativity—and the design of incentives—both social and economic—to support rather than hinder innovation. Creativity may be useless without adequate processes that support and nurture this creativity. Similarly, creativity and innovation can be damaged if incentives are counter-productive. Interestingly, the design of appropriate incentives varies with the type of innovation.

VOLUME 3: DESIGNING STRUCTURE AND SYSTEMS FOR SUPERIOR INNOVATION

The prior volumes deal with strategy and how to create an environment that encourages innovation. The focus of this third volume is how to design the organization and its management systems to support innovation. It addresses the third aspect that top management has to address in creating an innovative company: *designing the structures, processes, and systems that generate ideas, selecting the most promising ones, and transforming them into value.*

The volume also emphasizes the importance of cross-national interaction in getting innovation done. Three chapters address this issue from different perspectives. One of them examines the international component within product development. The second one looks at how venture capital—the money of innovation—has evolved from a regional to an international focus. Today, most venture capital firms' portfolios are diversified geographically with investments in North America, Europe, and Asia. A third chapter devotes its attention to how leading firms are managing R&D across borders. Different models are possible in addressing the need to coordinate knowledge from different parts of the world. But certain models are more adequate given the particular characteristics of the challenges at hand.

Another aspect relevant to the structure and systems of innovative organizations is the design of an appropriate measurement system. "What gets measured gets done" is frequently cited as a management principle, and it also applies to innovation management. But measures should not be used to evaluate performance, as they are sometimes used in other settings; their main role is to supply the information that guides discussion. Only in very specific types of innovation is it advisable to link measures to evaluation. Well-designed measurement systems track the entire innovation process. They provide information about the quality of the raw material for innovation—diversity of people, contact with the external world, and the quality of the ideas—all the way to the value created by innovation. In between, the system measures the balance of the innovation portfolio and the effectiveness of the innovation process.

Three chapters focus on organizing for innovation. One of them provides a balanced perspective between academic research and organizational applications on how to run product development projects. The second looks into the organization of novel ideas—usually harder to develop within an established organization—around the concept of incubators. Both chapters complement each other, providing the tools required to manage incremental and radical innovation. The third chapter presents the results of a research project on the characteristics of innovative firms. The study combines scientific rigor with enlightening examples.

An important issue in innovation management also addressed in this volume is intellectual property—in particular, how new intellectual property emerges from the combination of existing ideas. Innovation is not a blank page but the ability to combine existing ideas in novel ways.

Overall, the three volumes give a complete view of how to make an organization innovative. They balance depth in the state-of-the-art scientific knowledge with state-of-the-art managerial applications. We hope you will enjoy them!

Why Innovate? The Impact of Innovation on Firm Performance

DANIEL OYON

In recent years, innovation has become a central topic of discussion and thought in the business world. In an economy that has become more global and transparent through the deployment of new technologies for processing and distributing information, companies have put all their energy into being different, either by innovating in their business models—introducing new features in their products and services—or by modifying their processes to create a more efficient organization. Being different, whether it is through market positioning or execution, is critical to creating and capturing value. In the past, competitive advantage came from being better. Now it is also a question of being different.

This evolution did not just happen; rather, financial markets and investors drove it, increasing the amount of capital allocated to innovative companies. According to Gompers and Lerner (2001), between 1980 and 2000, venture capital investments in the United States increased more than a hundred times, from \$701.6 million to more than \$81,372.5 million. Capital has become less expensive and easier to mobilize for those managers who are trying to develop tomorrow's products and services. The creative destruction characteristics of a capitalist economy seem to be working more than ever at full steam: capital enters quickly those companies that can identify and secure privileged exploitation of future sources of value, and leaves just as fast out of companies competing in existing markets with no uniqueness.

Innovation is not only worshipped by financial markets. All the actors in the economy are interested. For instance, the media with its eternal fight for ratings has understood it and constantly emphasizes whatever is extraordinary. This would appear to be logical! Nothing grabs the attention of an audience as well as talking about new things, whatever they are.

Academic research has not remained on the sidelines; rather, it has been one of the leaders in this revolution. Scientists want dearly to push the limits of knowledge further than ever, and they are constantly searching for technological advances. Sociologists are tackling the issue of the extent to which technological progress leads to social progress. Economists are interested in understanding the conditions that allow the adoption of novelties that bring economic and social benefits. Lawyers are busy defining and introducing a legal framework that protects the rights of those who innovate.

In the management research field, innovation is also witnessing a significant investment of resources. For example, marketing researchers are looking to unravel the mysteries of consumer behavior and the evolution of their needs. Researchers in operations management are looking into innovating supply, production, and distribution processes to make them more effective and efficient for all the players involved. In organizational behavior, researchers seek to understand the elements that help creativity to blossom, ranging from organizational structures to culture to management systems. Last but not least, research on strategy has been interested for quite some time on product, process, and business model innovation as a critical source of competing advantage.

In spite of the considerable effort invested, many questions still remain unanswered, thus justifying continuous research in the field. This chapter addresses a fundamental but somewhat neglected question: does it pay to innovate? The first part of the chapter defines and introduces the various types of innovation that occur in the business world. The second part addresses the fundamental question of why it is important for an organization to innovate. The third part discusses why the innovation strategy of a company depends on its business objectives and its position in competitive markets.

WHAT DO WE MEAN BY INNOVATION?

Similar to other popular economic terms, the word "innovation" has a very important place in the discourse of political, economic, and social leaders. The fact that it is frequently alluded to not only highlights the importance of the topic but also generates much confusion, because its meaning depends strongly on the context in which it is used. Etymologically, innovation is the introduction of a new physical thing or a new method. In the business world, this definition refers to coming up with new product and service attributes that a company introduces to the market, and to novelties in the processes it executes. From a profitability point of view, companies look for innovations that increase the share of the value they capture—either through increased revenue, lower costs, or a better return on capital employed.

What type of innovation do we see being introduced in the market? Which companies are more innovative? Who benefits from innovation? What processes are needed to innovate? Plenty of relevant questions related to innovation exist, which explains the abundant literature relating to this topic. Many studies have addressed the question of how to classify innovation in terms of its importance to and impact on the market (Abernathy and Utterback 1978; Henderson and Clark 1990; Davila, Epstein, and Shelton 2005). One of the most common classifications is the one proposed by Markides and Geroski (2005). Four types of innovation are identified in Figure 1.1—incremental, major, radical, and strategic innovations—using a two-dimensional matrix that reflects the impact of innovation on both consumers and producers.

An innovation that has a minor impact on consumer habits and behavior and, at the same time, influences slightly capabilities and resources is an incremental innovation. New versions of the Intel Pentium processor or new versions of Microsoft Office are typical examples of incremental innovation. The set of attributes offered to the users is slightly extended. Companies doing this type of innovation develop their technical, commercial, and functional capabilities gradually. Incremental innovation follows a rhythm imposed by consumers that constantly demands new functionalities and better price-value ratio. This demand pushes companies to improve their products and services and to make their processes more efficient. Incremental innovations generally do not put the competitive position of existing companies in a given

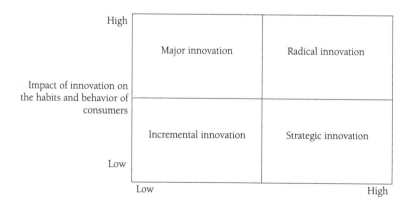

FIGURE 1.1. Different Types of Innovation according to Markides and Geroski (2005)

market at risk, except for companies that fail to follow the bandwagon and end up lagging behind the competition.

Sometimes innovation does not cause major changes of capabilities, but significantly transforms consumer habits and behavior. Markides and Geroski (2005) consider such innovations to be major innovations. In banking services, the arrival of automatic teller machines (ATMs) in the eighties led to significant changes in the way people dealt with banks. ATMs freed up customers from the constraint of having to go to a bank where they had their accounts during business hours to withdraw money; ATMs made this possible at any time of day or night. In the retail banking industry, this innovation was based on new technological and functional capabilities. These capabilities did not replace those necessary for managing traditional banking, because the latter remained essential in serving customers when certain transactions required providing advice and other services. ATMs were therefore a major innovation.

When innovation leads to major change in consumer habits and behavior and at the same time makes existing capabilities and resources suddenly obsolete, this is called radical innovation. The concepts of "breakthrough" and "discontinuous innovation" are also used to refer to this type of innovation. Ford launched its Model T in the early part of the last century, and IBM introduced the PC (personal computer) in the early eighties. Both are examples of radical innovation in the transportation and data processing industries. This type of innovation often ends a period of technology proliferation, leads to an emerging dominant architecture (dominating design), and redefines the competitive position of the different players. Indeed, the products and services associated with radical innovation displace the existing ones and destroy their markets; however, they provide a unique opportunity for those companies that introduce the dominant architecture to develop a major competitive advantage. In a short time, the configuration of the business model has swiftly changed, offering the companies championing the radical innovation the possibility to build an enviable competitive position.

Finally, certain innovations do not have a major impact on consumer habits and behavior, but cause dramatic changes in the competitive landscape. In this case, Markides and Geroski (2005) speak in terms of strategic innovation that leads to products and services being made available to new consumers. In contrast to radical innovation, this type of innovation is based not on new technologies but on setting up new business and organizational models. One of the most well-known examples of successful strategic innovation is Easyjet, an airline that carefully transposed and introduced the Southwest Airlines business model into the European context by offering point-to-point flights with a no-hub network. Another example is IKEA, the do-it-yourself furniture retailer that helps its customers visualize a wide range of furniture on site and then leaves it up to them to transport and assemble the goods.

WHY INNOVATE?

In spite of the abundant literature devoted to innovation, this fundamental question has seldom been addressed. Consequently, it has not received the attention that it deserves and remains an open question. The general public is often satisfied with categorical slogans, like "innovate or die," that transmit a binary image of the reality, where there are no alternatives because of the fear associated with death.

From a corporate perspective, this question is not only important, it is also the source of constant debate within organizations. While certain people adapt naturally to change and innovation, a large part of the population prefers the security associated with stability and lack of change. Therefore, understanding why innovation is so important for the survival of organizations is critical.

This question is even more relevant because innovating cannot be the *raison d'être* of an organization, even though mission statements sometimes reflect otherwise. In a market economy, the role of for-profit companies and not-for-profit organizations is clear. The former have to create and capture value for their owners and make it possible for the financial markets to quickly direct the capital toward the most promising sources of wealth creation. The latter must fulfill a mission of public interest, like providing assistance to people facing economic difficulties, offering access to health care, or promoting education and training programs. For both types of organizations, innovation is not an end in itself, but a means to accomplishing their goals.

Within this conceptual framework, we identify strategic, financial, commercial, organizational, and human reasons that make innovation essential to both the development and the longevity of companies.

The Strategic Reason: Innovate to Maintain or Improve the Competitive Position

Economists have always been interested in innovation from the perspective of competitive dynamics, which give certain companies incentives not to innovate while others have the opposite set of incentives. Sunk costs create a strategic asymmetry between established companies that have already invested in a certain business model, and new companies that have not yet committed themselves to a certain course of action. Following a rational decision process, an established company should not take into account the costs of its investments when evaluating new alternatives, because they are irreversible and unrecoverable (sunk costs). But actually, new opportunities are at a disadvantage against the status quo, and resistance to change is encouraged. This incentive deterring new investments is called the *sunk cost effect*. In contrast, a start-up usually has no firm commitments and can compare the various alternatives without being biased to maintain an existing business model.

The *sunk cost effect* is reinforced by the *replacement effect*, which stipulates that new entrants—due to their competitive position—have a greater incentive to find ways of displacing established firms from their market dominance. Arrow (1962) explains why an established company is less inclined to innovate or seek new competitive advantages than a start-up. To defend its argument, he examines the incentives involved in adopting a process innovation that sharply lowers variable manufacturing costs and makes the existing technology obsolete. He compares two different scenarios: (1) an established company has the opportunity to exploit the innovation and enjoy a monopoly through the technology, and (2) new and improved technology is available to a new company, which might achieve a monopoly position if it decides to adopt the technology. Arrow shows that the new company is willing to pay more than the established company to exploit the innovation. In fact, should it succeed, the new player would improve its competitive position significantly, while that of the established company would remain the same. Arrow concludes that new entrants will, under certain circumstances, replace monopolistic companies—not because the latter are poorly managed or are handicapped by agency problems, but because of the dynamics in the market that the starting positions of both companies create. This encourages newcomers to innovate and rewards established companies for blindly following the status quo.

The *efficiency effect* counterbalances the *replacement effect* and can encourage an established company that enjoys a monopoly situation to innovate in order to protect its competitive position and prevent the erosion of its profits. According to the well-known Cournot economic model, an established company that enjoys a monopoly has more to lose if a new company enters the industry than the new entrant has to win if it challenges the existing monopoly. The reason is that the arrival of a new entrant leads not only to sharing the market but also to lower prices. Therefore, the *efficiency effect* gives superior incentives to innovate compared to new companies.

Thus, innovation appears not only to be an end in itself but also a means to maintain or improve the competitive position. The different forces introduced above are present at the same time in any real situation. Whether one or the other will dominate a certain market will determine the competitive dynamics around innovation.

The Financial Reason: Innovate to Capture Additional Value

In a market economy, for-profit companies and not-for-profit organizations must constantly prove to fund providers that they are effective in using capital and in enhancing it. In this context, current and future financial performance plays a key role. The financial results of a for-profit company depend on the revenue from products and services that it sells in the market and on the costs of the resources consumed. Therefore, it is only possible to improve financial results in two ways: (1) increasing revenue and maintaining costs at a constant level or (2) maintaining revenue and reducing costs.

Revenue depends both on sales volume and on selling prices. In liberalized markets, four main elements affect selling prices: (1) the value created by the product or service, (2) the intensity of the competition among suppliers, (3) customer negotiating power, and (4) the available alternatives. Innovation in products and services can play an important role in financial results. Indeed, the value that the market determines and, consequently, the financial commitment that customers are ready to make to obtain this value depend directly on the tangible and intangible attributes of the product offered. Therefore, a new technology attribute is valued by its scarcity associated with novelty. Companies that bring such attributes to the market can expect to capture the value that they generate in the market.

However, capturing the value generated by innovation faces two main obstacles. The first obstacle is to correctly identify the new features that the market demands. This is not an easy task, even when the consumer indifference curve is increasing sharply—indicating a willingness to pay a higher price for a more attractive value—as illustrated in Figure 1.2. However, for most consumers, these innovations are generally of limited attractiveness because they consider that the current features of products and services satisfy their needs more than necessary (consumers whose indifference curve I1 is flatter beyond point P represent the value that the current product features offer). Thus, for these consumers, the new attributes will not translate into much higher prices—value that the company would capture. In contrast, for the small group of early adopters, the situation is different (consumers whose

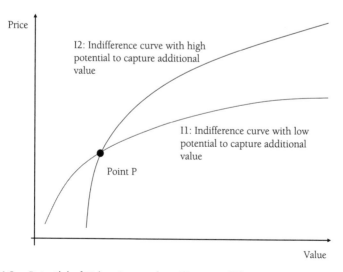

FIGURE 1.2. Potential of Value Capture for Different Indifference Curves

indifference curve I2 continues to strongly increase beyond point P). This type of consumer is willing to pay a much higher price for the new feature sometimes just for the pleasure of acquiring and possessing it. But identifying these early adopters and the attributes they value is not sufficient to capture the value of the innovation. This value can only be captured if a second obstacle is overcome, namely to produce these new features at a cost lower than the price that consumers are willing to pay. Due to the atmosphere of uncertainty created by innovation, costs are frequently much higher than expected and the value much lower.

The Commercial Reason: Innovate to Maintain and Grow Markets

Since the mid-1940s, marketing has taken a very relevant place in the world of business. It has moved from an era where customers asked producers to supply them with goods and services to satisfy their needs to a situation where producers have to approach existing markets to convince consumers to buy their products or create new markets that will attract consumers. Every day, whether for professional or personal reasons, people are bombarded with a vast array of messages through a diverse set of communication channels: commercial mail, publicity in the media, telemarketing, e-mail. Any and all means of getting messages through are utilized. In this space, attracting and maintaining the attention of customers has become a very difficult affair. It has become a game of carefully designing message content and form.

Managing to obtain an appointment to present a product or a service has become a feat. Consumers receive numerous requests to try new products at an increasing pace, and they need to set priorities to maximize their time and energy. In the modern business world, where it is difficult or impossible to be heeded, talking about innovation has become a way to develop and keep new relationships with existing and potential clients. Presenting a new feature or showing a new method for using a piece of equipment is a way to get an appointment. In such a context, innovation has become a means for opening new doors, attracting attention, satisfying curiosity, and avoiding relationships with customers becoming commonplace.

The Organizational Reason: Innovate to Learn

In highly competitive markets, people are under constant pressure to increase productivity. Nowadays, the motto "to make more with less (resources)" is embedded in all sectors, even in the luxury goods industry where, in theory, efficiency is not the underlying paradigm. Moreover, people, guided by their natural instincts work hard to move down the learning curve through specialization and repetition. Executing a task rapidly has always been socially and economically accepted. The pursuit of efficiency has led to a reduction in diversity of products and services and the dangerous spiral of standardization.

However, the environment is constantly changing: new technologies are appearing daily; consumer behavior has evolved along with changes in the socio-demographic structure of the population; the legal settings have also changed. Companies that do not adapt as quickly as the environment sooner or later experience major difficulties as their business models become obsolete. Certain companies, such as 3M, have understood well the need to adapt and are ready to sacrifice short-term profits to ensure long-term survival. Having put together true formal processes to generate and select various innovations, the famous Minnesota-based company forces products and services over a certain lifespan to be discontinued in order to stimulate the emergence of new products and to make room for them. Constant innovation allows skills, knowledge, and attitudes to be renewed permanently and remain up to date with the most stringent requirements of the environment.

The Human Factor Reason: Innovate to Attract and Retain Talent

In the twenty-first century, few industries are protected against the war of capabilities. In the knowledge economy, people are the true key factors of success. Without know-how and expertise, there is no economic health. Many companies have understood this dynamic well and have introduced advanced employee recruitment and selection policies to maintain and develop their organizational skills and knowledge.

In this unrestrained war for talent, innovation is an attractive proposition for such talent. Many talented people aspire to something other than working for a company. However, most people find fulfillment in work and try to test their skills and knowledge by joining companies where they will be able to stand out from the rest. Innovating through new products and services, changing management processes, or developing business models is a challenging task full of uncertainty, but a task that does not frighten those who are skilled and confident—rather, it attracts them and encourages them to perform. Influential companies such as Microsoft, Google, Cisco, McKinsey, or Goldman Sachs have understood what attracts talent and actively use innovation as an ingredient for recruitment.

WHAT INNOVATION IS NECESSARY FOR SURVIVAL?

In spite of the abundant literature on innovation, a normative model specifying the type of innovation that a company should pursue is still lacking. So far, the prevalent interest has been focused on identifying the sources of innovation and the reasons why established companies or new entrants have been, in some cases, better positioned to launch incremental or radical innovation (Abernathy and Utterback 1978; Henderson and Clark 1990). The conclusion is that the type of innovation depends mostly on the stock of competence and knowledge available or accessible, and on the fact that established firms have to respond to the needs of their existing clients rather than looking into new markets (Christensen 1997).

What type of innovation should a company pursue? This question worries many leaders because of both the financial and strategic resources that are at stake. Should companies let markets drive decision making and strive to meet the incremental needs that customers voice in order to keep their current positions? Should they seek to gain some ground by exploiting radical innovation? Should they survey today's very competitive markets from a distance and focus on tomorrow's markets through strategic or radical innovation to actively participate in their development and be well positioned to exploit them? Should they do both to take advantage of today's revenue for as long as possible, but at the risk of lacking focus and of being displaced by companies more focused on future markets?

Start-ups and Innovation

All innovation types are apparently possible for start-ups. Year after year, young companies are born with very diverse ambitions to take advantage of an immediate opportunity or to revolutionize the world. Although small companies created to meet the needs of a local segment of customers and to fulfill temporary deficiencies within the market play an important role in society, capital markets are more interested in start-ups focused on radical or strategic innovations that may create large new markets and capture large amounts of value. However, the failure rate of radical and strategic innovations is very high. In their book *Fast Second*, Markides and Geroski (2005) document in detail the difficulties that a pioneer company faces. The number of companies that try to innovate radically is huge compared to those that succeed. Creating a dominant architecture (dominating design) in industries such as automobile, aviation, computer, software, or recorders requires efforts that span many years and many organizations, from research centers with government funding to private companies. As depicted in Figure 1.3, the emergence of a radical innovation like the browser and the Web search engine materialized by the IPO of Netscape or Google is the "result of the 'crusade.'" Many start-ups fell along the wayside throughout the journey. However, the collective efforts in this "crusade" have been decisive in the process toward an emergent dominant architecture. According to Markides and Geroski (2005), radical innovation seldom originates in existing markets that are too focused on tracking the evolution of their customers' needs; rather, it stems from the particular skills and knowledge developed in research centers that technology promoters help move into value. It also originates from the vision of pioneers like Frederick Smith, founder of Federal Express, who can project themselves into the future and mobilize resources over long periods of time to move through the various

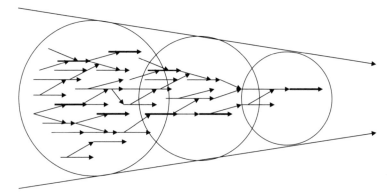

FIGURE 1.3. Trajectory of a Dominant Architecture

stages. However, these pioneers rarely obtain a strategic and financial reward commensurate to their efforts (Olleros 1986), even if the exploration process is undoubtedly an enriching experience from an intellectual and human perspective to which pioneers are sensitive.

Established Companies and Innovation

It is very difficult and even unwise for established companies to play an active role in the phases preceding the emergence of radical or strategic innovation. The competencies, knowledge, and management systems in place do not allow them to have the right perspective on their existing markets. They are not aligned to tomorrow's markets, which could emerge after a disruption caused by new technology, new legal rules, or a change in spending patterns. For these companies, financial logic prevails: *a dollar today is worth more than one dollar tomorrow, and a sure dollar is worth more than an uncertain one.* Satisfying the needs of their existing customers is often the only objective for which structures, systems, and culture have been designed. Established companies are fitted for incremental innovation in products and services, advancing with the evolution of consumer behavior and their product and service requirements. The objective of process innovation is to maintain the price-cost ratio, offering more value to the customer and increasing the share of the value generated. The competition between established companies is often intense because each one tries to offer better value while optimizing the price-value ratio. In this game, the key to success is to offer additional value at a cost that is fully covered by the increase in price. But disappointments are not foreign to incremental innovation. However, a strategy of incremental innovation is hardly dangerous in the short term because consumers provide fast feedback and companies can quickly react if they do not meet consumers' needs precisely. For example, in early 2006, the Nestlé group decided to fully rejuvenate its famous Cailler chocolate line by redesigning the brand through the introduction of new packaging and the update of its recipes to adapt them to the evolving tastes of consumers. In spite of the importance of the amount invested, this initiative is an incremental innovation because it did not question Nestlé's basic competences in chocolate. Although Nestlé did not publicly report the strategic and financial results of this innovation, they could hardly have been extraordinary because the company did not increase its market share substantially or capture much value of the Swiss consumers already over-served in terms of chocolate products.

However, pursuing incremental innovation is not always strategically and financially neutral. A study of incremental innovation launched in the car industry shows that the most innovative companies have not obtained adequate rewards for their efforts. Indeed, Table 1.1 below represents the distribution of the incremental innovations made between 1995 and 2005 and indicates that the innovation leadership at Mercedes apparently did not translate into value for its shareholders, whose stock performance over those ten years had been negative. Moreover, it is interesting to note that Porsche, in spite of launching few innovations, generated a substantial stock return. However, it is difficult to take conclusive lessons from this table since, on the one hand, stock price is the outcome of a number of factors, not only of the degree of innovation, and, on the other hand, certain innovations, although incremental, can be more important than others in terms of impact on the market.

Regarding radical innovation, Markides and Geroski (2005) discuss profusely the life cycle of this type of innovation and the respective roles of the start-up companies (pioneers) and the established companies (consolidators). It seems that established companies are better off leaving the pioneer role to

TABLE 1.1. Incremental Innovations and Stock Market Performance

Company	% of innovations after 1995	Sales	Net Profit	Annual stock market performance
Mercedes	21.74%	$149,976	$5,185	−1.20%
BMW	16.30%	$46,656	$2,239	23.48%
Volkswagen Group	15.22%	$95,268	$1,120	13.52%
Peugeot Citroen	7.61%	$68,686	$1,220	16.67%
Renault-Nissan	7.61%	$49,606	$4,144	26.05%
Toyota	6.52%	$172,749	$10,907	18.34%
Ford	5.43%	$153,503	$2,024	−4.67%
Porsche	4.35%	$7,887	$934	175.87%
Honda	2.17%	$29,579	$1,224	22.96%

research centers and small specialized companies and focusing on actively managing a portfolio of real options in order to be able to exploit the opportunities offered by an emerging dominant architecture and the blossoming of new markets. As Figure 1.4 shows, creating new markets out of radical technologies goes through stages beginning at research centers, then start-ups, and finally established companies. Radical innovations that do not follow this order often lead to situations of strategic and financial distress, both for start-ups and for established companies. Because the mental model of established companies is centered on their existing markets, their structures, their resource allocation processes, and their performance evaluation systems, these companies are not prepared to pursue radical innovation or to carry out all the stages leading to an emerging dominant architecture.

There are many examples of established companies that have suffered dearly, notably Xerox, a company that makes innovation its *raison d'être* and not a means to financial and strategic performance. It is interesting to note that its research center in Palo Alto, CA (PARC), inspired an impressive number of radical innovations that did not lead to strong strategic and financial results for the company. Xerox built the first portable computer in 1973, long before the birth of Apple Computers and more than eight years before the launch of the IBM PC. The Xerox research center also developed the mouse as a pointing device and human-machine interface. Both innovations formed the foundation for the subsequent success of Apple Macintosh and Microsoft Windows. Another interesting fact is that Xerox invented word processing several years before Microsoft even existed; it also created the first laser printer, the local area network, and the first object-oriented

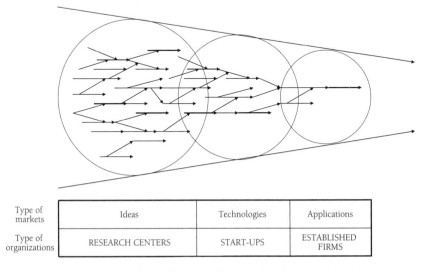

Type of markets	Ideas	Technologies	Applications
Type of organizations	RESEARCH CENTERS	START-UPS	ESTABLISHED FIRMS

FIGURE 1.4. Dominant Design Trajectory and Roles of Various Organizations

programming language. Despite these revolutionary technologies, for which Xerox is unanimously recognized (as the creator), the company never benefited from the value they created and did not build any strategic and financial advantage, even though it supported most of the research and development costs. Over this period of time, Xerox became a renowned company. However, its market value of $13 billion as of April 2006 has remained a lot lower than that of companies like Microsoft ($242 billion), Intel ($115 billion), and Dell ($57 billion), which are better known as consolidators than as radical pioneers.

The unfortunate case of Xerox has shown established companies that their competitive advantage does not reside in the development of radical new technologies but in large-scale exploitation of technologies invented by others. Companies like Cisco and Pfizer have realized that they need to focus on commercializing these innovations rather than on conducting the initial development phases of any of these technologies. Therefore, Pfizer's remarkable stock price performance in the first years of the twenty-first century was mainly explained by its ability to market Lipitor, a cholesterol-fighting drug developed by Warner Lambert. The invention of Viagra played a role, but not to the same extent. However, it is not easy to recognize the emergence of a radical innovation and be ready for it. Much like IBM's founder, who thought that the size of the market for large computers was limited to a handful of potential customers, the examples of pioneers who did not see the potential of promising radical innovations such as the telephone, the computer, or the Internet are more numerous than those who were aware of the long-term implications of the emerging dominant architecture.

CONCLUSION

In a world of continuous change, innovation is essential for any type of business, whether it is a for-profit or a not-for-profit organization. The goal of the former is protecting and developing the value captured in the exchange of products and services that takes place through increasingly competitive markets. For the latter, innovation enables them to increase in efficiency and thus protects their *raison d'être*.

There are four types of innovation in product and services markets, on one side incremental and major innovations and on the other side radical and strategic innovations. Incremental innovations are important for existing markets since they facilitate constant adaptation to evolving consumer needs. But incremental innovation does not redistribute the existing market position of the various players, and thus they are not a significant source for a long-lasting competitive advantage. In contrast, radical and strategic innovations can ruin existing companies and allow new entrants to thrive. In this context, there is nothing to lose with incremental innovation, and much to gain with radical innovation.

To mobilize long-term capital, new or established companies must prove that they are able to succeed in the long run and thus cannot ignore radical innovation. However, they must be aware of their individual and specific roles in the market if they want to succeed. Indeed, the role of research centers is to build the foundation of radical innovations. In contrast, start-ups should take the baton when they observe strong signals that ideas could have a potential market value. Established firms should enter the game when the emerging market is ready to be scaled up based on the success of a dominant design. At this point of the twenty-first century, public financing, venture capital, and capital markets are clearly structured with this view and look for the best companies to develop ideas, transform them into exploitable technology, and then scale them for a large market. Failing to recognize this sequence can lead to serious business and financial failures.

REFERENCES

Abernathy, W. and Utterback, J. (1978) Patterns of industrial innovation. *Technology Review*, 80 (7): 40–47.

Arrow, K. (1962) Economics welfare and the allocation of resources for inventions. In R. Nelson (Ed.), *The rate and direction of inventive activity*. Princeton, NJ: Princeton University Press.

Christensen, C. (1997) *The innovator's dilemma: When new technologies cause great firms to fail*. Boston: Harvard Business School Press.

Davila, A., Epstein, M. and Shelton, R. (2005) *Making innovation work*. Philadelphia: Wharton Business School.

Gompers, P. and Lerner, J. (2001) *The money of invention*. Boston: Harvard Business School Press.

Henderson, R. and Clark, K. (1990) Architectural innovation: The reconfiguration of existing product technologies and the failure of established firms. *Administrative Science Quarterly*, 35 (1): 9–30.

Markides, C. and Geroski, P. (2005) *Fast second*. San Francisco: Jossey-Bass.

Olleros, F. (1986) Emerging industries and the burnout of pioneers. *Journal of Product Innovation Management*, March, 5–18.

2

On Dynamic Clustering, Innovation, and the Role of IT

RAMON O'CALLAGHAN

The advantages of industrial clusters with respect to standalone firms are well known. Already in 1920, Alfred Marshall developed the concept of "external economies of scale" to refer to sources of productivity that lie outside of individual firms, e.g., sharing fixed costs of common resources, such as infrastructure and services, skilled labor pools, specialized suppliers, and a common knowledge base. When these factors are geographically concentrated, firms gain the benefits of spatial proximity. But in the twenty-first century, one can posit that the new way of clustering is based on knowledge and competencies, and not geographical proximity and the relative "inward looking" nature of traditional clusters and industrial districts, perhaps best typified by the Italian experience (Becattini 1979).

This chapter argues the need to research emerging forms of "virtual" or "extended" clusters, i.e., those that transcend location, focus on international markets, operate as ad-hoc business networks, are IT-enabled, and are based on dynamic aggregation of capabilities of different (often small) firms. The working hypothesis is that these new organizational arrangements, which in this chapter are called "extended dynamic clustering" (EDC), can help small companies position themselves better in terms of global market access and innovation.

The chapter is structured as follows: First, it reviews some relevant literature on the underlying concepts and issues, e.g., IT-enabled business transformation, knowledge and learning in clusters, and dynamic capabilities. Then it develops a conceptual framework and research agenda on dynamic clustering and networks. Finally, it discusses potential policy implications of the proposed research.

IT, ORGANIZATIONAL CHANGE, AND BUSINESS PERFORMANCE: BEYOND THE PRODUCTIVITY PARADOX

Research on the link between organization design and business performance has a long tradition (e.g., Chandler 1962; Thompson 1967; Galbraith 1977; Caves 1980; Quinn 1980; Porter 1985). In the past decade, the role of IT as enabler of organizational design and organizational transformation has become a topic of interest in both the information systems literature as well as the general management literature (Hammer 1990; Scott-Morton 1991; Davenport 1993; Hammer and Champy 1993). By redesigning the way existing business processes are performed and using IT to enable new ones, some organizations have been able to achieve significant improvements in key business drivers, such as cost, quality, service levels, or lead times.

Yet these successes did not seem to make an impact on productivity figures at the macroeconomic level. Robert Solow's (1987) famous quip that "You can see the computer age everywhere but in the productivity statistics" provoked a great deal of debate. A similar debate has emerged more recently, regarding the value of IT for competitive advantage with some authors claiming that IT is a commodity (Carr 2003). Yet the suggestion that IT does not bring benefits to organizations seems to go against intuition and common sense. If IT investments do not yield any clear advantages, why do so many organizations continue to invest heavily in IT?

Subsequent research has tried to explain away the "IT productivity paradox" (e.g., Brynjolfsson and Hitt 1998; Willcocks and Lester 1999; Triplett 1999). Pilat and Wyckoff (2004), as well as Brynjolfsson and Hitt (2004), show that the use of IT is positively linked to firm performance. Other studies reveal substantial differences between organizations that utilize IT in a successful versus an unsuccessful way (Brynjolfsson and Hitt 1998).

The IT productivity paradox should not be a disquieting problem for managers. After all, there seem to be many opportunities for individual organizations to use IT in innovative and profitable ways. The question for managers is not whether IT pays off but what IT applications they should deploy in their respective organizations.

Over the years, the focuses of IT application and IT management have been shifting from efficiency-related issues to the question of how to deliver business value with IT. In the early days, IT had merely automating effects, which could easily be justified on the basis of cost savings, but today's investments have "transforming" effects, e.g., improving quality, flexibility, and the innovation ability of organizations. Paradoxically, it is this transformational power that makes it difficult to pinpoint the exact contribution of IT to business value. IT is so widespread, and so intertwined with business processes, that it cannot be looked at in isolation. In this respect, two major areas of IT application have been particularly significant in recent years: systems for knowledge management and interorganizational systems in the supply chain. The next sections describe them in more detail.

Knowledge Management for Innovation

Knowledge Management (KM) encompasses processes and practices concerned with the creation, acquisition, capture, sharing, and use of knowledge, skills, and expertise (Quintas, Anderson, and Finkelstein 1996). The KM literature has traditionally focused on IT and information systems (IS) to create network structures that can link together individuals distributed across time and space. Many articles have focused on developing and implementing KM databases, tools, and techniques for the creation of "knowledge bases," "knowledge webs," and "knowledge exchanges" (Bank 1996).

Behind this approach to KM lies an information-processing view of the firm where valuable knowledge located inside people's heads is identified, captured, and processed through the use of IT tools so that it can be applied in new contexts. Workers' knowledge is thus captured and made accessible to others via a search engine (Cole-Gomolski 1997). The idea is to reduce problems of "reinventing the wheel" by "exploiting" existing knowledge more efficiently, i.e., by deploying it in other similar situations.

KM, however, should not be confined to "exploitation." "Exploration" (i.e., where knowledge is shared and synthesized and new knowledge is created) is more important for innovation (Levinthal and March 1993). It is exploration through knowledge sharing that allows the development of genuinely new approaches (Swan et al. 1999).

Increasingly, knowledge is being produced interactively at the point of application among heterogeneous groups (Gibbons 1994). Innovation processes are thus becoming more interactive—more dependent on knowledge that is widely distributed. Organizational design trends are aligned with this focus on KM for innovation. The new organizations are typified by flatter structures, decentralization, collaboration, and coordination through use of information and communication technologies. However, as businesses are stretched across time and space, reorganized along process or product lines, and restructured around virtual teams, they lose opportunities for innovation through the casual sharing of knowledge and learning induced by physical proximity (Swan et al. 1999).

KM for interactive innovation has implications for the deployment of IT as well as for the management of people and social networks. While early generations of knowledge management solutions focused on explicit knowledge in the form of documents and databases, there is a need today to expand the scope of the solutions to include technologies that can support tacit knowledge (Marwick 2001). Future applications will then have to address the following needs: assist teams in sharing experiences, build and share tacit knowledge; help groups work effectively together and support collaboration; conduct electronic meetings and trust building through (video) conferencing; identify individuals with the right knowledge; elicit assistance from experts and the community; support the formation of new tacit knowledge from explicit knowledge (portals, taxonomies, knowledge mapping, etc.).

Transcending Organizational Boundaries

The other major area of IT application with significant implications for enterprise transformation involves interorganizational systems, e.g., collaborative systems in supply chains. Typified by the cases of some leading companies (e.g., Wal-Mart and Dell, among others), these systems show that the locus of change and innovation is no longer confined within the boundaries of the organization. Some of the most dramatic changes have taken place at the level of supply chains or business networks, as exemplified by the case of the PC industry.

Since the mid-1990s, the PC industry has been using direct sales channels, demand-driven production, and modular production networks. Within these networks, firms are flexible in designing value chains for different products and markets, with each firm selecting a different mix that takes into account its own capabilities and strategies. The structure of the industry's global production network changed, making it possible to coordinate design, production, and logistics on a regional or global basis (Kraemer and Dedrick 2004). As a result, PC makers have been able to locate these activities where costs are low and key skills are available, or else close to major markets. Also, the use of IT, the Internet, and e-commerce have enabled and supported the shift from supply-driven to demand-driven production and the creation of more flexible, information-intensive value chains to support this complex process. This change has led to dramatic reductions in inventory, better use of assets, and leaner operations throughout the industry.

The sources of competitive advantage in the new IT-enabled organization are the substitution of information for inventory, better matching of supply and demand, and the ability to tap into external economies in the global production network. External economies can be accessed by any firm, but demand-driven organizations are best positioned to take advantage of these economies because they can use real-time information to drive the production network in response to demand changes.

New models of outsourced manufacturing (e.g., contract manufacturing and manufacturing services) emerged also in globalized production networks in the electronics industry (Lüthje 2004). These models have been enabled by the interaction of information networks as well as by the restructuring of production work and the global division of labor. Information technology is not the driver of organizational change per se, but part of a complex shift in the social division of labor that ultimately is related to the demise of vertically integrated mass manufacturing. In this context, information technology and Internet-based models of supply-chain management do facilitate vertical specialization. An important issue is the question of network governance. In this respect, a relevant trend is the centralization of supply-chain management in electronic components. So, the issue is how to orchestrate complex networks of corporate actors and their interaction in global marketplaces.

The shift from supply-driven to demand-driven value chains has had important effects not only in industrial markets but also in the area of consumer interactions. The "reversal" of the value chain, together with the Internet, empowers consumers in ways that were unimaginable just a few years ago. Consumers today can create virtual communities and engage in an active dialogue with manufacturers of products and services. At the same time, consumers constitute a source of knowledge that companies can exploit. This transforms the traditional notion of "core competence" (Prahalad and Hamel 1990). Competence now becomes a function of the collective knowledge available in the ecosystem, i.e., an enhanced network comprising the company, its suppliers, its distributors, its customers, its partners, and its partners' suppliers and customers.

In this customer-centric approach, firms are no longer producers of products or services but (co)developers of customer experiences. Companies engage customers in an active, explicit, and ongoing dialogue, mobilize consumer communities, manage customer diversity, and co-create personalized experiences with customers (Prahalad and Ramaswamy 2000). Organizations that can "sense and respond" rapidly by moving information to mobilize resources and knowledge in the network are expected to emerge as the "winners" in the network economy (Bradley and Nolan 1998; Kraemer and Dedrick 2004). But if mobilizing remote resources on a network is the way of the future, what will happen to the advantages traditionally associated with physical proximity (e.g., in industrial clusters)?

Location Matters: Regional Clusters

Much has been made of the potential of IT to enable a despatialization of economic activity. Cairncross (1997), among others, posits that with the introduction of the Internet and new communications technologies, distance as a relevant factor in the conduct of business is becoming irrelevant. She contends that the "death of distance" will be the single most important economic force shaping all of society over the next half century.

Despite the bold predictions, however, geography and location still matter. Porter's identification of local agglomerations, based on a large-scale empirical analysis of the internationally competitive industries for several countries, has been especially influential, and his term "industrial cluster" has become the standard concept in this field (Porter 1998, 2001). Also, the work of Krugman

(1991, 1996) has been concerned with the economic theory of the spatial localization of industry. Both authors have argued that the economic geography of a
nation is key to understanding its growth and international competitiveness.

Alfred Marshall developed the concept of "external economies of scale"
(Marshall 1920) to refer to sources of productivity that lie outside of individual firms, e.g., sharing fixed costs of common resources, such as infrastructure and services, skilled labor pools, specialized suppliers, and a common
knowledge base. When those factors are geographically concentrated, firms
gain the benefits of spatial proximity (Storper 1989).

Regional clusters are examples of external economies derived from industrial localization. They are self-reinforcing agglomerations of technical skill,
venture capital, specialized suppliers, infrastructure, and spillovers of knowledge associated with proximity to universities and informal information flows
(Hall and Markusen 1985; Arthur, 1990).

Other researchers see regional economies as networks of relationships rather
than as clusters of individual firms. The network approach offers insights into
the structure and dynamics of regional economies by focusing on the relationships between firms and the social structures and institutions of their particular
localities (Powell 1990; Nohria and Eccles 1992). This view has been used to
explain the divergent trajectories of Silicon Valley and Boston's Route 128
economies (Saxenian 1994).

Regional clusters can also be viewed as "complex systems." Complexity
theory focuses on the study of emergent order in what are otherwise very dis-
orderly systems that are neither centrally planned nor centrally controlled
(Holland 1998; Anderson et al. 1999).

Complex systems innovate by producing spontaneous, systemic bouts of
novelty out of which new patterns of behavior emerge. Patterns that enhance
a system's ability to adapt successfully to its environment are stabilized and
repeated (single loop learning); those that do not are rejected in favor of radically new ones, often by trial and error (double loop learning) (Argyris 1997).

A systemic view of clusters must therefore distinguish between two possible regimes: "directed order" and "emergent order" (or simply "order" and
"un-order") (Kurtz and Snowden 2003). The ordered system is governed by a
series of established "routines" that have emerged over time through repeated
actions and interactions of the clustered firms. From a knowledge management perspective, this regime can be characterized as "exploitation" (March
1991), given that the cumulated knowledge (embedded in the rules and routines) is applied to deal with known problems. In contrast, the regime of "un-
order" deals with new problems for which the solution is not known. This regime involves trial-and-error experiments because the established routines and
knowledge are no longer valid. From a knowledge perspective, this regime
can be called "exploration," as new knowledge must be acquired or created to
deal with the new, more turbulent environment (March 1991).

In practice, however, exploration and exploitation co-exist. The term "reconfigurability" has been used by some researchers to focus on revamping existing functional competencies versus destroying them entirely in favor of new ones (Grant 1996). Their views are in line with the theory of dynamic capabilities that stresses the notion of achieving new configurations (Teece, Pisano, and Shuen 1997) to adapt quickly to the changing needs in the environment. Reconfiguration is also particularly relevant in new product development, where most new products are inventive recombinations of existing competencies that better match customer needs (Henderson and Clark 1990). The notion of "reconfigurability" can also be extended to clusters, as will be discussed in a subsequent section of this chapter.

Network Governance: The Missing Link

Miles and Snow (1986) introduced their view of enterprise networks as a flexible, fluctuating and dynamic structure. The concept of business networks gained more attention after publication of the bestseller *The Virtual Corporation* by Davidow and Malone (1992). Today numerous network typologies can be found in the literature (Powell 1990). Proposals range from strategic hub-and-satellite networks, as in the automotive industry to clan-like structures, as in Japanese Keiretsus (Ouchi 1980), and regional networks up to temporary networks and dynamic virtual organizations.

Most publications on business networks have in common that they are predominantly descriptive. The models depict possible emerging outcomes of network structures, loose couplings, and collaboration among companies. Findings on efficient management and controlling procedures can hardly be deducted from these models. Furthermore, the majority of the research has focused on the general characteristics of organically evolved networks, and on their structure and development processes. Much less attention has been paid to intentionally developed nets and their management, with the notable exceptions of the work of Jarillo (1993) and Parolini (1999) on value nets, and the emerging theory of network governance in economic sociology and strategic management (Amit and Zott 2001; Gulati, Nohria, and Zaheer 2000; Jones, Hesterly, and Borgatti 1997). Thus, the challenges involved in operating in a complex network remain fairly unarticulated.

BEYOND CLUSTERS: SMALL FIRMS AMIDST GLOBALIZATION

The trend toward globalization of the economy poses a number of challenges to smaller firms in traditional clusters. Often, due to size, scale, specialization, and, not least, regulatory and legal impediments, small and medium enterprises (SMEs) lack the capacity to respond adequately to market opportunities or participate in tenders in international procurement contracts. This shortcoming is related to both the conditions that SMEs face and the operation of geographically based clusters. More specifically, one can distinguish "internal" reasons (specific to the SMEs) and "external" reasons (specific to clusters and insufficiently developed cross-border and cross-regional collaboration mechanisms among clusters):

- Internal reasons have to do with limited resources and competences. SMEs often do not possess all the relevant skills and competencies, and cannot afford the specialized human resources (e.g., legal and technical expertise) required to participate in collaborative cross-border or cross-region processes for the co-creation and delivery of products and services;
- External reasons span from the perceived complexities of international contract negotiation, to trust and financial issues, as well as the perceived disadvantages in terms of size and skills (e.g., SMEs may rule themselves out when they know that some large competitors will be bidding). External reasons include also regulatory and legal gaps that create roadblocks to cross-border collaboration, contract negotiation, intra- and inter-cluster governance policy and institutional issues that hinder the formation and efficient operation of cross-border and cross-regional collaborative networks.

From these two perspectives, a fundamental challenge is how to facilitate linkages, not only among SMEs within a given cluster but also across clusters and networks of SMEs. This challenge involves building "internal" capabilities by enhancing the organizational knowledge and technological capacity of SMEs to enter into cross-border and cross-regional collaborative processes for jointly producing and delivering products and services. It also involves building "external" capacity in the environments in which SMEs and their clusters operate. In other words, if the "internal" set of issues refers to the business challenges SMEs face, the "external" issues concern the "enabling framework" that will facilitate cross-border and cross-regional collaboration among SME clusters.

TOWARD A FRAMEWORK FOR EXTENDED DYNAMIC CLUSTERING

As argued above, the challenge for SMEs is to create the conditions for "extended and dynamic clustering" based on the notion of identifying and selecting complementary resources and capabilities out of SME networks that "extend" beyond the boundaries of a traditional cluster (e.g., regional or international instead of just a local pool of potential collaboration partners). Thus, the working hypothesis is that "extended dynamic clustering" (EDC) can help small companies position themselves better in terms of global market access and innovation.

A NEW PARADIGM: EXTENDED DYNAMIC CLUSTERS

"Extended" clusters are conceptualized as virtual clusters that transcend location, focus on international markets, operate as ad-hoc business networks, are IT-enabled, and are based on a dynamic aggregation of capabilities of different (often small) firms. "Dynamic" clusters aggregate SMEs from different industrial sectors, involved in different processes and also operating in different markets. The advantage is that the resulting "extended dynamic" cluster is much more responsive and enjoys a steep learning curve. The knowledge base and the competence mix in the dynamic cluster determine the speed and level of the response as well as the necessary structural changes (e.g., industrial culture, internal and external processes, relationships) many of which could not be possible to achieve for an SME operating in a standalone way.

An important question is how little changes inside the cluster (e.g., changing or adding a key new partner) can bring significant changes in the ability to respond to opportunities in the market. This involves a knowledge transfer process. Let's consider, for example, a cluster specialized in producing mechanical parts and tools for the automotive sector. They decide to respond to a tender from an aerospace company, and, because they lack some necessary skills, they decide to include in the cluster a supplier operating in the aerospace sector. The added competence of this new partner gives the cluster the possibility not only to go in the new marketplace, but to learn "by immersion" in a new industrial environment. This "full immersion learning" is learning not only from the new partner, but also from all the players in the aerospace environment, i.e., customers, competitors, suppliers, the regulatory agency, etc. Thus, in a short period of time, the cluster learns and evolves into a "new" type of cluster that now can operate in a new sector. Repeating this process several times improves the dynamic capabilities and thus the flexibility of the cluster to innovate, incorporate new technologies, and tackle new markets.

One way to understand the notion of "extended dynamic" clustering is by positioning this new construct against traditional forms of business agglomeration, e.g., industrial clusters and business networks. The diagram in Figure 2.1 shows the two dimensions that characterize this evolved cluster form.

The horizontal dimension is based on the typology found in the literature on business networks that differentiates "dynamic" from "static" networks (Miles and Snow 1986). The vertical dimension represents the geographic scope (operations space) of a given "virtual" or "extended" cluster. This dimension can be operationalized essentially as the geographic distance between the collaborative

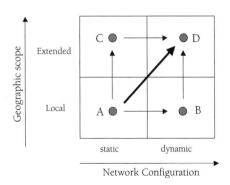

FIGURE 2.1. Clustering Typology

firms. In practice, it may be useful to differentiate between local, regional, national, and transnational domains. This differentiation is particularly important for governance. The governance issues and potential policy recommendations are likely to differ at local, regional, national, and supranational levels.

PATH ANALYSIS: BARRIERS AND ENABLERS

As indicated above, the working hypothesis is that type D clustering is superior to types A, B, or C in the sense that product innovation and "market access" are improved. If D is the desired state, then the questions can be phrased in terms of the enablers/barriers required to move from type A to type D clustering. As the diagram suggests, a differential analysis can be performed along the different paths. Such analysis will require the identification and assessment of a sufficiently large number of networks (with various degrees of variance along the two dimensions) so as to populate all the quadrants of the matrix. Enablers that facilitate collaboration when geographic spread increases can be analyzed along the following paths:

- Path A–C: focus on the networks operating in a *static* mode and compare the barriers and enablers of local/regional networks with those of interregional and transnational networks.
- Path B–D: focus on the networks operating in a *dynamic* mode and compare the barriers and enablers of local/regional networks with those of interregional and transnational networks.

Enablers that facilitate dynamic interaction and reconfiguration can be analyzed along these paths:

- Path A–B: focus on *local/regional* networks and compare the barriers and enablers of networks operating in a static mode with those operating in a dynamic mode.
- Path C–D: focus on *transregional* and *transnational* networks and compare the barriers and enablers of networks operating in a static mode with those operating in a dynamic mode.

Dynamic Capabilities for Clustering

The strategic management literature has traditionally focused on analyses of firm-level strategies for sustaining and protecting extant competitive advantage, but has performed less well with respect to assisting in the understanding of how and why certain firms build competitive advantage in regimes of rapid change. To address this problem, researchers have focused on "dynamic capabilities" (Teece, Pisano, and Shuen 1997). Dynamic capabilities are defined as the "ability to integrate, build, and reconfigure internal and external competences to address rapidly changing environments."

Dynamic capabilities should not be confused with functional competencies. Functional competencies are purposive combinations of resources that enable tasks or operational activities (e.g., logistics and manufacturing). Dynamic capabilities, on the other hand, are the abilities to revamp functional competencies. Thus, researchers distinguish between dynamic capabilities that connote change (first-order), and functional (zero-order) competencies (King and Tucci 2002; Winter 2003). Similarly, Henderson and Clark (1990) distinguish between component innovation and architectural innovation. The former is based on functional competencies, and the latter is based on knowing how functional competencies are integrated and linked together into a coherent whole.

The application of the dynamic capabilities framework to cluster-level analysis would suggest that the competitive advantage of a SME cluster rests on the distinctive processes for coordinating and combining capabilities, shaped by the cluster's (unique) asset positions (such as its portfolio of difficult-to-trade knowledge assets) and the evolution path(s) it has adopted or inherited.

Reconfigurability, Innovation, and Market Access

Reconfiguration is generally viewed as the ultimate outcome of dynamic capabilities. Most studies in the dynamic capabilities literature stress the importance of reconfiguring existing resources into new configurations of functional competencies. For example, reconfigurability refers to the timeliness (Zott 2003) and efficiency (Kogut and Zander 1996) by which existing resources can be reconfigured (Galunic and Rodan 1998). It refers also to the concept of "combinative capabilities" (Kogut and Zander 1992) that describes the novel synthesis of existing resources into new applications. Eisenhardt and Brown (1999) introduced the term "patching" to reflect the ability to "quickly reconfigure resources into the right chunks at the right scale to address shifting market opportunities."

While dynamic capabilities can reconfigure all resources (Prahalad and Ramaswamy 2004), it is important to stress the role of knowledge as an intangible resource (Galunic and Rodan 1998; Glazer 1991). Leonard-Barton (1992) argues that as resources become less tangible, visible, and explicitly codified, they will be easier to reconfigure.

Following the knowledge-based view (Grant 1996), reconfiguring knowledge into new knowledge sets can develop productive new competencies. Dynamic capabilities thus reflect "the ability to learn new domains" (Danneels 2002). Hence, their value lies in the configurations of functional competencies they create (Eisenhardt and Martin 2000; Zott, 2003). For example, by spotting market trends and accordingly revamping functional competencies, dynamic capabilities can prevent rigidities (Leonard-Barton 1992) and competency traps (March 1991). Also, by replacing outdated configurations of functional competencies and architecting more relevant ones, dynamic capabilities can create better matches between the new configurations of functional competencies and environmental conditions (Teece, Pisano, and Shuen 1997).

Applied to extended clusters, dynamic capabilities would enable SMEs networks to redeploy their existing (outdated) competencies to build new products or services through innovative, aggregated competencies that better match emerging market and technological needs.

Potential Enablers

A distinction must be made between the reconfiguration itself (i.e., deployment) and the enabling processes that facilitate reconfiguration (Pavlou and El-Sawy 2005). The dynamic capabilities and related literatures describe four processes that drive reconfiguration, innovation, and change:

1. *Sensing the environment:* Sensing helps understand the environment, identify market needs, and spot new opportunities (Zollo and Winter 2002).
2. *Learning:* Learning builds new thinking, generates new knowledge, and enhances existing resources (Zollo and Winter 2002).
3. *Coordinating Activities:* Coordinating helps allocate resources, assign tasks, and synchronize activities (Teece, Pisano, and Shuen 1997).
4. *Integrating Resources:* Integrating resources helps implement the new architectural innovations by developing the patterns of interaction (Grant 1996; Henderson and Clark 1990).

Because of its abstract nature, the concept of reconfigurability is difficult to assess directly. The above four tangible enabling processes, however, can be potentially operationalized and measured, thereby overcoming the criticism that dynamic capabilities do not consist of specific, identifiable, and concrete processes. Pavlou and El-Sawy (2005) propose a set of literature-driven constructs for the above enabling processes:

- Sensing the Environment is captured by the construct of *"Market Orientation"* (Kohli and Jaworski 1990),
- Learning by *"Absorptive Capacity"* (Cohen and Levinthal 1990),
- Coordinating Activities by *"Coordination Capability"* (Malone and Crowston 1994),
- Integrating Resources by *"Collective Mind"* (Weick and Roberts 1993).

These constructs originate from different literatures and have been studied at different levels and units of analysis. The methodological challenge is to adapt and extend these constructs to units of analysis appropriate for cluster/business network research.

The Role of Information Technology

In traditional clusters, the need for physical proximity has led to regional agglomerations. Clusters have thus depended on face-to-face contacts. But relying exclusively on physical proximity limits the available talent pool and the access to specialized facilities. So there is a strong case for taking advantage of IT to link to remote professionals and resources.

Innovative uses of information and communications technology enable a "despatialization" of economic activity and, at the same time, offer new opportunities for codifying information, which may enhance learning and innovative activity. Future research should look at clusters become geographically proximate complex organizational systems of learning and economic and social activity that are globally networked and enabled by the effective use of IT. Some relevant questions are:

- How will IT affect traditionally perceived needs for physical proximity and introduce "virtual" proximity as a complement to physical proximity?
- Can "virtual" clusters be expected to emerge and/or develop, in part, as a result of the widespread application of IT?
- What combinations of physically proximate and "virtual" arrangements best augment the social and economic performance of networked clusters?

One way to address these questions is by focusing on the enablers of extended dynamic clustering identified above. The following sections put forward specific issues regarding potential roles of information technology to enable clustering capabilities along the two dimensions identified in Figure 2.2, i.e., virtual proximity capabilities and dynamic clustering capabilities.

IT, Distance, and Virtual Proximity

Some researchers argue that knowledge cannot be shared or absorbed independently of the processes through which it is generated (Roberts, 2000). But, if greater stocks of knowledge can be circulated across electronic networks and used in ways that effectively support learning, then the importance of geographical clustering and physical presence may indeed be reduced.

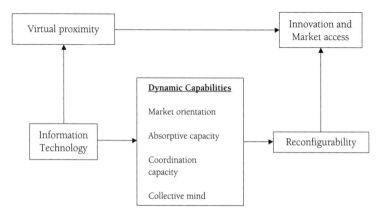

FIGURE 2.2. Role of IT in Extended Dynamic Clustering

Figure 2.3 shows a relationship between the degree of codification of knowledge and the speed and extent of its diffusion within a target population (O'Callaghan and Andreu 2006). In other words, the figure highlights a tradeoff between codification and reach. The shape of the curve indicates that more people can be reached per unit of time with knowledge that is codified (explicit) than with knowledge that remains uncodified (tacit). As the size of the target population that one seeks to reach increases, the message needs to be more highly codified to reach that population quickly, and much of the contextual richness of the message must be sacrificed for the sake of communicative efficiency.

New IT applications can change the nature of this tradeoff between loss of context and speed of diffusion. By increasing data processing and transmission capacities, they enable more data to reach more people, whatever the level of codification chosen, as indicated in Figure 2.3. This is shown by a horizontal shift in the curve.

The horizontal arrow shows how at a given level of codification, the population to which a message can be diffused increases. But the downward-pointing vertical arrow shows something else: it suggests that, for a given size of population being targeted, a message can be sent at a lower level of codification than in the absence of IT, i.e., the message can transmit more of its context, thus restoring some of the context-specific interpersonal qualities usually sacrificed to codification (e.g., videoconferencing).

The figure also reinforces the need, discussed above, to expand the scope of IT solutions in knowledge management, to include technologies that can support tacit knowledge, assist teams share experiences, help groups work

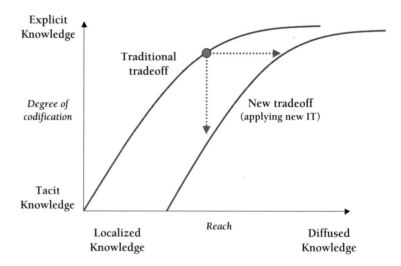

FIGURE 2.3. Knowledge Flows and IT Impacts

effectively together, etc. The use of diverse IT solutions within and between clusters is likely to have implications for the meaning of proximity.

The following paragraphs focus on the potential role of IT to enable or support dynamic clustering capabilities. The discussion is structured around the four constructs identified above: market orientation, absorptive capacity, coordination, and collective mind.

IT for Market Orientation

Market orientation reflects the ability to sense the environment and under-stand customer needs and competitive dynamics. It is defined as "the process of generating, disseminating, and responding to market intelligence about customer needs" (Jaworski and Kohli 1993; Kohli and Jaworski 1990). The relevant questions for the role of IT are the following:

- How can IT be used for *capturing market intelligence*, e.g., external communication links for sensing market trends or discovering new market opportunities?
- How can IT be used for *disseminating market intelligence* to the appropriate parties in the business network/virtual cluster?
- How can IT be used to *analyze and interpret* market intelligence?
- How can IT help *respond to market trends*, e.g., by enabling processes and supporting operations that capitalize on market intelligence?

Some of these market-oriented dynamic capabilities are best exemplified by the use of information technology to enable operations at Zara, the clothing retailer and manufacturer described in the Harvard Business Case "Zara: IT for Fast Fashion" (McAffee, Dessain, and Sjoman 2004). Zara is able to respond very quickly to the demands of young, fashion-conscious city dwellers whose tastes in clothing change rapidly and are hard to predict and influence. Every Zara store places an order to headquarters in La Coruna, Spain, twice a week. The order encompasses both replenishments of existing items and initial requests for newly available garments. Managers learn about newly available garments by consulting a handheld computer that is linked each night, via a dial-up modem, to information systems at La Coruna. A digital order form with suggestions (called "the offer") is transmitted to all stores. The offer includes descriptions and pictures of newly available items, as well as replenishments. Each store's offer is different. Offers are developed by a team of marketing specialists based on garment availability, regional sales pattern, predictions about what will sell in each location, and other factors.

In contrast to other large clothing retailers, Zara not only introduces new collections at the start of every season but also brings out new items continuously throughout the year. Zara's vertically integrated manufacturing enables this constant introduction of items and also ensures short lead times.

Production requirements are distributed across a network of specialized facilities that quickly produce and deliver the required goods. Zara owns a group of factories in and around La Coruna to do the capital-intensive initial production steps of dyeing and cutting cloth. Cut fabric is sewn into garments at a network of small local workshops in Galicia and northern Portugal that guarantee quick turnaround times. Using this network, Zara can consistently move a new design from conception through production and into a distribution center in as little as three weeks. Two days later, the garments can be on sales racks in stores around the world. This speed enables Zara to respond to the fast-changing and unpredictable tastes of its target customers. No other apparel retailer can match this capability. This speed translates into a high rate of innovation. Zara introduces 11,000 new items every year, whereas competitors average 2,000 to 4,000 items.

While the Zara case illustrates market-oriented dynamic capabilities mostly at the organizational level, the research challenge put forward in this chapter is to construct and operationalize these market-oriented capabilities at the interorganizational and network levels. With respect to technology's role, the larger question is how to deploy IT, both internally and interorganizationally, as an effective platform for market intelligence gathering, analysis, and dissemination in the distributed environment inherent in extended dynamic clusters.

IT for Absorptive Capacity

Absorptive capacity reflects the ability to learn by identifying, assimilating, transforming, and exploiting existing knowledge resources to generate new knowledge (Cohen and Levinthal 1990; Zahra and George 2002). Regarding IT, the relevant research questions are:

- How can IT help acquire or "broker" knowledge?
- How can IT help assimilate knowledge (e.g., through knowledge articulation and codification)?
- How can IT help transform knowledge, (e.g., in generating new thinking, brainstorming and experimentation, innovative problem solving)?
- How can IT help exploit knowledge (e.g., in pursuing new initiatives and identifying new solutions)?

The use of IT for learning and knowledge management is exemplified by Buckman Laboratories (Fulmer 1999). Buckman is a leading manufacturer of specialty chemicals for water industrial systems. For many years the company had been sending out its Ph.D.s to gather best business practices worldwide and then share with all associates in the company. The problem was they could not hire enough Ph.D.s and run them fast enough to do face-to-face exchanges around the world. Then a formal system to capture and share knowledge within the company was started, using a database to record how sales associates created new knowledge within the organization. New knowledge was solving a problem at a customer's business either by applying "existing Buckman knowledge" or by developing a new, more effective or efficient solution if it was a new problem.

Over time, the system evolved to include the "tacit" knowledge of the associates of the organization. The idea was to connect people through a network that would "replace the multi-tiered hierarchy with the breadth of knowledge that is the collective experience of employees." This type of knowledge transfer system, which Buckman named K'Netix, was based on forums: "open places" where anyone can post a message, question, and/or request for help. All the messages relating to a particular topic are collected together as a thread and filed in dedicated areas within a library. System operators were appointed to monitor the discussions in the forums, track requests, and make sure they were answered.

K'Netix became a new way of operating the business. The system allowed the company to change its strategy from "selling chemicals" to "selling solutions" ("We don't sell chemicals; we sell problem-solving expertise."). Buckman attributes much of the 250 percent sales growth in the past decade to knowledge sharing. Likewise, the high rate of innovation (35 percent of sales come from products fewer than five years old) is also attributed to knowledge sharing.

IT for Coordination

Coordination capability reflects the ability to manage dependencies among resources and tasks to create new ways of performing a desired set of activities (Crowston 1997; Malone and Crowston 1994). Pertinent questions are:

- How can IT be used to allocate resources (including distribution of knowledge)?
- How can IT help assign tasks among partners?
- How can IT be used to appoint the right person to the right unit?
- How can IT help synchronize activities among collaborating partners?
- How can IT be used to capture synergies among tasks and resources?

IT for "Collective Mind"

"Collective mind" has been defined as the "ability to integrate disparate inputs through heedful contribution, representation, and subordination into a group system" (Weick and Roberts 1993). It can also be conceptualized as the architecture for the whole system. In this respect, "collective mind" helps implement a set of complex activities by specifying the organizing principles by which individual knowledge is integrated (Grant 1996). The IT-related questions are:

- How can IT be used to model and structure the cluster/network?
- How can IT be used to monitor how partners fit in and interact, and how their activities affect others?
- How can IT be used to interrelate diverse inputs (including knowledge) from constituent firms to execute the collective activity of the cluster/network?

- How can IT help individual inputs contribute to the group outcome? How can IT support the sharing of knowledge among partners?
- How can IT be used to keep network managers informed?

An example of how IT is used both for coordination of interorganizational activities and for "collective mind" is the holistic supply change management of Li & Fung, the Hong Kong import-export trading company (Young 2000). Li & Fung (Trading) Limited is a premier global trading group managing the supply chain for high-volume, time-sensitive consumer goods. As a supply chain manager across many producers and countries, Li & Fung provides the convenience of a one-stop shop for customers: from product design and development, through raw material and factory sourcing, production planning and management, quality assurance and export documentation to shipping consolidation.

Li & Fung provides value-added services across the entire supply chain in a "borderless" manufacturing environment. For example, a down jacket's filling might come from China, the outer shell fabric from Korea, the zippers from Japan, the inner lining from Taiwan, and the elastics, label, Velcro, and other trim from Hong Kong. The garment might be dyed in South Asia, stitched in China, sent back to Hong Kong for quality control, and then finally packaged for delivery to retailers in the U.S. or Europe. As an executive of Li & Fung puts it: "we do not own any of the boxes in the supply chain, rather we manage and orchestrate it from above. The creation of value is based on a holistic conception of the value chain" (Young 2000). When clients place an order, Li & Fung determines the manufacturers best suited to supply the goods. In addition, Li & Fung orders the raw materials and offers them to the manufacturers, ensuring both greater quality control and economies of scale, and therefore saving costs for each manufacturer. Effectively, the company customizes the value chain to best meet the customer's needs.

Li & Fung's IT includes secure extranet sites to link the company directly to key customers. These sites are customized to the customers' individual needs. Through these sites, Li & Fung can carry out online product development as well as order tracking, obviating costs and time associated with sending hard copies back and forth. Furthermore, with Li & Fung as the key link between manufacturers and retailers, the extranet provides a platform for streamlining communications in the supply chain. Customers can track orders online. This monitoring of production also promotes quick-response manufacturing. Until the fabric is dyed, the customer can change the color; until the fabric is cut, the customer can change the styles or sizes offered, whether a pocket or a cuff can be added, and a number of other specifications. Some customers connect their ERP systems to Li & Fung's extranet system.

GOING FORWARD

The working hypothesis is that market access will be improved through a double process of "extended" and "dynamic" clustering. The "extended" clustering process (EDC) implies selecting and aggregating capabilities of clustered SMEs at the regional, national, or international level, thus overcoming the local geographical boundaries and operational limitations of traditional clusters. The "dynamic" clustering process implies flexibility and adaptability in aggregating and configuring "virtual" clusters for the purposes of responding to specific and changing market opportunities. The goal is to understand the mechanisms that enable, facilitate or impede such processes. In essence, future knowledge will develop around the following steps:

1. Develop a *conceptual framework* that provides the theoretical foundation and the tools for empirical analysis of the above-mentioned concepts (i.e., dynamic clustering).
2. Test the hypothesis that *dynamic clustering facilitates market access for SMEs.*
3. Establish and assess the *enablers and barriers* (economic, social, technological, legal, institutional, and policy-related factors) that facilitate or impede extended dynamic clustering,
4. Analyze the role that IT can play in lowering the barriers identified in the preceding point. For example, how an open, decentralized networking environment can enable SMEs to build cross-national virtual consortia to supply integrated services in response to specific market opportunities.
5. Derive policy implications.

Potential research byproducts resulting from the above agenda might include:

- Conditions for an outward-looking perspective on clusters with emphasis on the traditional economic strengths of regions but also on dynamic capabilities to respond to rapid economic changes and global competition.
- Conditions for reconfiguring clusters as "hubs" and roles of institutions in helping build regional economic capacity (in terms of dynamic capabilities, networking, and international connections) to enable regional SMEs to confront the challenges of being "hubs" between a global economy and a regional business ecosystem.
- How SMEs have to reconfigure themselves from being simple members of a local cluster to being important nodes of a global network of suppliers and retail outlets.
- IT tools to enable the process of finding and selecting the appropriate partners to respond to a given market opportunity, e.g., tools for locating and aggregating "expertise" and other resources.
- The effects of open-source IT platforms and tools that may support new methods of collaboration, and process integration within, between, and across regional networks incorporating SMEs and large contracting organizations, as part of an end-to-end supply chain.

- Regulatory/policy impediments and enabling conditions for cross-border and cross-regional dynamic cluster formation and collaboration.
- Regulatory, legal, organizational and technological impediments to the "pulling power," i.e., the capacity to attract investment and innovation activities, of clusters with emphasis on the conditions favoring synergies and collaboration between and across them.
- Domains for policy intervention in terms of regulation, legal measures, technology policy at supranational, national, and regional levels for the creation and facilitation of dynamic clustering.

RELEVANCE AND POLICY IMPLICATIONS

Over the past few years, the cluster concept has found a ready audience among policymakers at all levels, from the World Bank, to national governments, to regional development bodies, to city authorities. All are keen to find a new form of industrial policy in which the focus is firmly on the promotion of successful competitive economies. The argument is not that governments can create clusters, but that they can help foster the business, innovative, and institutional environments vital for their success.

The first step is usually to identify the clusters in the region or country. Organizations and government agencies that view their regions as clustered production systems are predisposed to tailoring existing policies and programs to that model and in some instances creating new strategies. The most common policy levers are those that alter the way agencies organize and deliver their services, work with employers, recruit businesses, and allocate resources. But the most popular goals are to market a political region and attract businesses and highly educated and skilled people.

A key question that the EDC paradigm raises is what it means to market a region when traditional clusters morph into extended dynamic clusters and the relative importance of the territory diminishes. Understanding the processes and factors shaping dynamic clustering in a region will allow policymakers to adapt policies and programs to the "extended dynamic cluster" model and, in some cases, formulate new regional development strategies.

Social and Economic Impacts

The focus is not only on the interregional and transnational dynamic clustering and competition, taken together, may foster economic development by improving the capacity of SMEs to innovate and access (global) markets.

One of the factors often associated with the success of traditional clusters is social capital. Social networks expose members to new processes and markets, nonpublic bid requests, and innovations. However, companies outside the networks may miss out on many economic opportunities. Clusters create a capacity to network and learn, but they are often characterized by intangible barriers to entry. In conventional clusters, access to the learning network may be controlled by the interests of some large companies. This has traditionally been the problem for SMEs, which, as a result, have been slow to learn about and adopt new technologies, or to enter new markets. One of the propositions is that extended dynamic clustering may lead to greater social inclusion. The rationale for such a proposition is based on the premise that knowledge in extended dynamic clusters is more freely available and is not limited to the local resources controlled by a few.

A number of regions classified as "less favored" have sectors specialized in traditional industries with little innovation and predominance of small family firms with weak links to external markets. The most successful clusters, on the other hand, include lead firms that are parts of global networks and are exposed to global market opportunities, and that employ people active in international professional associations and networks. These firms regularly benchmark themselves against the best practices anywhere. Poorer regions and smaller companies have limited access to benchmark practices, innovations, and markets. Without wider access, companies are limited to learning only within their regional borders and have a difficult time achieving any sort of competitive position. Future research should ascertain whether, and how, extended dynamic clustering can increase the social inclusion of poorer regions and their SMEs.

Policy Implications

The EDC paradigm can provide a new lens for policy research and practice. To apply the EDC concept to policy, one must believe not just that extended dynamic clustering is possible, but also that it makes a difference and that it can be influenced.

It needs to be explored whether the EDC framework allows policymakers to identify more accurately market imperfections, find pressure points, envisage or identify systemic failures, and determine what interventions can have the greatest impacts. Are new policies not required when regions are examined from an EDC perspective?"

Extended dynamic clusters differ from traditional clusters in their extraterritorial reach, dynamic capabilities, and the enabling role of IT. Information technologies provide a new means of linking up local places and regions within networks of organizations. Inclusion in the network requires an adequate local technological infrastructure, a system of ancillary firms and other organizations providing support services, a specialized labor market, and a system of services required by the professional labor force. Research outcomes should include guidelines for policymakers and civil society organizations in order to facilitate the transitioning of SMEs to extended dynamic clusters, as well as the adoption and usage of related ITs.

The "new industrial spaces" of today are comprised of complex networks with more than one central node. They can be seen as geographically proximate, complex organizational systems of learning and economic activity that are globally networked with other systems. The spread of global, national, regional, and local IT networks and information flows may fuel an "innovative milieu" (cross-regional and transnational) and act as a catalyst for social learning processes that give rise to successful and enduring economic and social development. If public policymakers proactively encourage the integration of advanced ITs and their application to link geographically clustered firms with other organizations beyond their immediate regional surroundings, there may be substantial opportunities for a departure from the conventional pattern of regional development and a catalyst for growth.

REFERENCES

Amit, R. and Zott, C. (2001) Value creation in e-business. *Strategic Management Journal*, 22: 493–520.

Anderson, P., Meyer, A., Eisenhardt, K., Carley, K., and Pettigrew, A. (1999) Introduction to the special issue: Applications of complexity theory to organizational science. *Organization Science*, 10: 3.

Argyris, C. (1977) Double loop learning in organizations. *Harvard Business Review* (September/October): 115–125.

Arthur, B. (1990) Positive feedbacks in the economy. *Scientific American*, 262 (2): 92–99.

Becattini, G. (1979) Dal settore industriale al distretto industriale. *Rivista di Economia e Politica Industriale*, n. 1.

Bank, D. (1996) Technology-know-it-alls—chief knowledge officers have a crucial job: putting the collective knowledge of a company at every worker's fingertips. *Wall Street Journal, Eastern edition*. November 18: 28.

Bradley, S. P. and Nolan, R. L., eds. (1998) *Sense and respond: Capturing value in the network era*. Boston: Harvard Business School Press.

Brynjolfsson, E. and Hitt, L. M. (1998) Beyond the productivity paradox: Computers are the catalyst for bigger changes. *Communications of the ACM*, 41 (8): 49–55.

Brynjolfsson, E. and Hitt, L. M. (2004). Intangible assets and the economic impact of computers. In Dutton, W., Kahin, B., O'Callaghan, R., and Wyckoff, A., eds. *Transforming enterprise*. Cambridge, Mass.: MIT Press. 27–48.

Cairncross, F. (1997) *Death of distance: How the communications revolution will change our lives and our work*. Boston: Harvard Business School Press.

Carr, N. (2003) Does IT matter? *Harvard Business Review* (May): 5–12.

Caves, R. (1980) Industrial organization, corporate strategy and structure. *Journal of Economic Literature*, 18: 64–92.

Chandler, A. D. (1962) *Strategy and structure: Chapters in the history of the American industrial enterprise*. Cambridge, Mass.: MIT Press.

Cohen, W. M. and Levinthal, D. A. (1990) Absorptive capacity: A new perspective on learning and innovation. *Administrative Science Quarterly*, 35: 128–152.

Cole-Gomolski, B. (1997) Users loath to share their know-how. *Computerworld*, 31 (46): 6.

Crowston, K. (1997) A coordination theory approach to organizational process design. *Organization Science*, 8 (2): 157–175.

Danneels, E. (2002) The dynamic of product innovation and firm competences. *Strategic Management Journal*, 23 (9): 1095–1121.

Davenport, T. H. (1993) *Process innovation: Reengineering work through information technology.* Boston: Harvard Business School Press.

Davidow, W. H. and Malone, M. S. (1992) *The virtual corporation: Structuring and revitalizing the corporation for the 21st century.* New York: HarperCollins.

Eisenhardt, K. and Brown, S. (1992) Patching: Restitching business portfolios in dynamic markets. *Harvard Business Review,* 77: 72–82.

Eisenhardt, K. and Martin, J. (2000) Dynamic capabilities: What are they? *Strategic Management Journal,* 21: 1105–1121.

Fulmer, W. E. (1999) Buckman laboratories (A). *Harvard Business School case no. 9-800-160.* Boston: Harvard Business School Press.

Galbraith, J. R. (1977) Organizational design. Boston: Addison-Wesley.

Galunic, D. C. and Rodan, S. (1998) Resource recombinations in the firm: Knowledge structures and the potential for Schumpeterian innovation. *Strategic Management Journal,* 19: 1193–1201.

Gibbons, M. (1994) *The new production of knowledge: The dynamics of science and research in contemporary societies.* London: Sage Publications, Ltd.

Gulati, R., Nohria, N. and Zaheer, A. (2000) Guest editors' introduction to the special issue: Strategic networks. *Strategic Management Journal,* 21: 199–201.

Glazer, R. (1991) Marketing in an information-intensive environment: Strategic implications of knowledge as an asset. *Journal of Marketing,* 55: 1–19.

Grant, R. (1996) Toward a knowledge based theory of the firm. *Strategic Management Journal,* 17: 109–122.

Hall, P. and Markusen, A. (1985) *Silicon landscapes.* Boston: Allen & Unwin.

Hammer, M. (1990). "Reengineering work: Don't automate, obliterate." *Harvard Business Review* (July/August): 104–113.

Hammer, M. and Champy, J. (1993). *Reengineering the corporation: A manifesto for business revolution.* New York: HarperBusiness.

Henderson, J. C. and Clark, H. (1990) Architectural innovation. *Administrative Science Quarterly,* 35: 9–30.

Holland, J. (1998) *Emergence: From chaos to order.* Boston: Addison-Wesley.

Jarillo, J. C. (1993) *Strategic networks: Creating the borderless organization.* Bodmin, Cornwall: MPG Books, Ltd.

Jaworski, B. J. and Kohli, A. (1993) Market orientation: Antecedents and consequences. *Journal of Marketing,* 57: 53–70.

Jones, C., Hesterly, W. S., and Borgatti, S. P. (1997) A general theory of network governance: Exchange conditions and social mechanisms. *Academy of Management Review,* 22 (4): 911–945.

King, A. and Tucci, C. (2002) Incumbent entry into new market niches: The role of experience and managerial choice in the creation of dynamic capabilities. *Management Science,* 48 (2): 171–186.

Kogut, B. and Zander, U. (1992) Knowledge of the firm, combinative capabilities, and the replication of technology. *Organization Science,* 3 (3): 383–397.

————. (1996). What firms do? Coordination, identity, and learning. *Organization Science,* 7 (5): 502–518.

Kohli, A. K. and Jaworski, B. J. (1990) Market orientation: The construct, research propositions, and managerial implications. *Journal of Marketing,* 54: 1–18.

Kraemer, K. L., and Dedrick, J. (2004) The role of information technology in transforming the personal computer industry. In Dutton, W., Kahin, B., O'Callaghan, R., and Wyckoff, A., eds. *Transforming enterprise*. Cambridge, Mass.: MIT Press: 313–334.

Krugman, P. (1991) *Geography and trade*. Cambridge, Mass.: MIT Press.

Krugman, P. (1996). The localisation of the global economy. In Krugman P., ed. *Pop Internationalism*. Cambridge, Mass.: MIT Press.

Kurtz, C. F. and Snowden, D. J. (2003) The new dynamics of strategy: Sense-making in a complex and complicated world. *IBM Systems Journal*, 42 (3): 462–483.

Leonard-Barton, D. (1992) Core capabilities and core rigidities: A paradox in managing new product development. *Strategic Management Journal*, 13 (Summer Special Issue): 111–125.

Levinthal, D. and March, J. (1993) The myopia of learning. *Strategic Management Journal*, 14: 95–112.

Lüthje, B. (2004) IT and the changing social division of labor: The case of electronics contract manufacturing. In Dutton, W., Kahin, B., O'Callaghan, R., and Wyckoff, A., eds. *Transforming enterprise*. Cambridge, Mass.: MIT Press 2004: 335–358.

Malone, T. and Crowston, K. (1994) The interdisciplinary study of coordination. *ACM Computing Surveys*, 26 (1): 87–119.

March, J. (1991) Exploration and exploitation in organizational learning. *Organization Science*, 2: 71–87.

Marshall, A. (1920) *Industry and trade*. London: Macmillan.

Marwick, A. D. (2001) Knowledge management technology. *IBM Systems Journal*, 40 (4): 814–830.

McAffee, A., Dessain, V., and Sjoman, A. (2004) Zara: IT for fast fashion. *Harvard Business School case no. 9-604-081*. Boston: Harvard Business School Press.

Miles, R. E. and Snow, C. C. (1986) Organizations: New concepts for new forms. *California Management Review*, 28: 62–72.

Nohria, N. and Eccles, R., eds. (1992) *Networks and organizations: Structure, form, and action*. Boston: Harvard Business School Press.

O'Callaghan, R. and Andreu, R. (2006) "Knowledge dynamics in regional economies: A research framework. In Sprague, R., ed. *Proceedings of the 39th HICSS (Hawaii International Conference on System Sciences)*. Washington, D.C.: IEEE Computer Press.

Ouchi, William G. (1980) Markets, bureaucracies, and clans. *Administrative Science Quarterly*, 25: 129–141.

Parolini, C. (1999) *The value net: A tool for competitive strategy*. Chichester, England: John Wiley & Sons, Ltd.

Pavlou, P. A. and El Sawy, O. (2005) Understanding the "black box" of dynamic capabilities. *Management Science* (under third round of review).

Pilat, D. and Wyckoff, A. W. (2004) The impacts of IT on economic performance: An international comparison at three levels of analysis. In Dutton, W., Kahin, B., O'Callaghan, R., and Wyckoff, A., eds. *Transforming enterprise*. Cambridge, Mass.: MIT Press: 77–110.

Porter, M. (1985) *Competitive advantage: Creating and sustaining superior performance*. New York: The Free Press.

————. (1998) Clusters and the new economics of competition. *Harvard Business Review*, 76 (6): 77–90.

————. (2001). *Clusters of innovation: Regional foundations of U.S. competitiveness*. Washington, D.C.: Council on Competitiveness.

Powell, W. W. (1990) Neither market nor hierarchy: Network forms of organization. *Research in Organizational Behavior*, 12: 295–336.

Prahalad, C. K. and Hamel, G. (1990) The core competence in the corporation. *Harvard Business Review* (November/December): 79–91.

Prahalad, C. K. and Ramaswamy, V. (2000) Co-opting customer competence. *Harvard Business Review* (January/February): 79.

Prahalad, C. and Ramaswamy, V. (2004) *The future of competition*. Cambridge, Mass.: Harvard Business School Press.

Quinn, J. B. (1980) *Strategies for change: Logical incrementalism*. Homewood, Ill.: Irwin.

Quintas, J. B., Anderson, P., and Finkelstein, S. (1996) Managing professional intellect: Making the most of the best. *Harvard Business Review*, 74 (March/April): 71–80.

Roberts, J. (2000) From know-how to show-how: Questioning the role of information and communication technologies in knowledge transfer. *Technology Analysis and Strategic Management*, 12 (4): 429–443.

Saxenian, A. (1994) *Regional advantage: Culture and competition in Silicon Valley and Route 128*. Cambridge, Mass.: Harvard University Press.

Scott-Morton, M. S., ed. (1991) *The Corporation of the 1990s: Information technology and organizational transformation*. Oxford: Oxford University Press.

Solow, R. M. (1987) We'd better watch out. *New York Times Book Review* (July 12): 36.

Storper, M. (1989). The transition to flexible specialization in the U.S. film industry: External economies, division of labor, and the crossing of the industrial divides. *Cambridge Journal of Economics*, 13: 273–305.

Swan, J., Newell, S., Scarbrough, H., and Hislop, D. (1999) Knowledge management and innovation: Networks and networking. *Journal of Knowledge Management*, 3 (4): 262.

Teece, D. J., Pisano, G., and Shuen, A. (1997) Dynamic capabilities and strategic management. *Strategic Management Journal*, 18 (7): 509–533.

Thompson, J. D. (1967) *Organizations in action*. New York: McGraw-Hill.

Triplett, J. E. (1999) The Solow productivity paradox: What do computers do to productivity? *Canadian Journal of Economics*, 32 (2): 309–334.

Weick, K. E. and Roberts, K. H. (1993) Collective mind in organizations: Heedful interrelating on flight decks. *Administrative Science Quarterly*, 38 (3): 357–381.

Weill, P. (1990) *Do computers pay off? A study of information technology investment and manufacturing performance*. Washington, D.C.: ICIT Press.

Willcocks, L. and Lester, S., eds. (1999) *Beyond the IT productivity paradox*. Chichester, England: John Wiley and Sons.

Winter, S. (2003) Understanding dynamic capabilities. *Strategic Management Journal*, 24 (10): 991–995.

Young, F. (2000) Li & Fung. *Harvard Business School case no. 9-301-009*, Boston: Harvard Business School Press.

Zahra, S. A. and George, G. (2002) Absorptive capacity: A review, reconceptualization, and extension. *Academy of Management Review*, 27 (2): 185–203.

Zollo, M. and Winter, S. G. (2002) Deliberate learning and the evolution of dynamic capabilities. *Organization Science*, 13: 339–351.

Zott, C. (2003) Dynamic capabilities and the emergence of intra-industry differential firm performance: Insights from a simulation study. *Strategic Management Journal*, 24: 97–125.

3

Toward a Non-linear History of R&D: Examples from American Industry, 1870–1970

W. BERNARD CARLSON

F or the linear model of R&D, it was the best of times and the worst of times. In the late 1950s, the president of the Corning Glass Works, Bill Decker, had remarked to the head of R&D, William Armistead, "Glass breaks ... why don't you fix that?" In response to this challenge, Armistead had his scientists investigate all known ways of strengthening glass. Drawing on theory and experiments, Corning scientists developed a new chemical treatment known as ion exchange that could be applied to glass after it had been formed. Remarkably, this new chemically strengthened glass could withstand pressures up to 100,000 pounds per square inch, as compared to 7,000 pounds per square inch for ordinary glass. Christened Chemcor, this new glass was a triumph of scientific research, and Corning proudly announced its discovery with great fanfare in 1962.

Yet unlike DuPont's nylon of the 1930s, Chemcor did not become the wonder material of the 1960s. While Corning researchers had been able to come up with a new glass, the company had not identified any uses or customers for Chemcor. (As we will see below, this was a departure from a long

The material in this chapter on Corning Incorporated is based on research and writing I have done as a consultant to Corning. I am grateful to Stuart Sammis of Corning for advice and information. Naturally, any opinions or conclusions drawn here are my own and not those of the management of Corning. A version of this chapter was presented at the Conference on Science and Technology in the 20th Century: Cultures of Innovation in Germany and the United States German Historical Institute, Washington, D.C., October 2004.

tradition in which Corning scientists and managers worked together to link discoveries with markets.) Instead, at the press conference announcing Chemcor, Corning invited the public to suggest potential applications. Over the next several years, Corning fielded thousands of suggestions, but only a few made it into production, and none became a major product.

Corning subsequently used Chemcor to make automobile windshields. To make windshields out of Chemcor, the company had to develop a new manufacturing technique, the fusion draw process. However, after spending millions of dollars, Corning learned that the major automakers in Detroit were not interested in changing from the existing safety glass to Chemcor windshields. In part, this was because the automakers made windshields in their own glass plants. It was also because the Chemcor windshield would have cost slightly more than the existing safety glass, and the automakers would not gain any benefit by offering Chemcor windshields to their customers. By 1969, Corning had spent over $42 million on Chemcor and had still not converted this idea into a profitable business.[1] (Notably, Corning now employs the fusion draw process to make LCD glass for the flat screens in computer monitors and televisions, which has proven to be a very lucrative business.)

The story of Chemcor reveals that, while science can generate wondrous new things, these new things do not necessarily translate automatically into new products, profits, or jobs. Yet, since the end of World War II, American leaders in science, business, and government have assumed that science, industry, and economic growth are governed by a simple set of relationships. Science provides the theory that business applies to create new products, and new products generate prosperity and jobs. Known as the linear model, these relationships were articulated by Vannevar Bush in his 1945 report *Science: The Endless Frontier*. After noting how science had helped win the war through the rapid development of penicillin and radar, Bush argued that "What we often forget are the millions of pay envelopes on a peace-time Saturday night which are filled because new products and new industries have provided jobs for countless Americans. Science made that possible too."[2]

Over the years, historians have generally assumed that R&D evolved in a straightforward linear fashion in the late nineteenth and early twentieth centuries; as scientists developed new theories about electricity and chemistry, it was inevitable that businessmen would exploit this new knowledge. Moreover, as businessmen created larger corporations, they had the money to invest in research. And, of course, it was not problematic for the managers of business firms to integrate science into their organizations, since science surely equaled profits. This interpretation has a strong flavor of determinism, in the sense that scientific and technological change—in the forms of electricity and chemistry—brought on inevitable social change in the form of the new R&D laboratory.[3]

In this chapter, I want to suggest that there was no linear evolution of the relationship of science and industry in the American context. From the 1870s to the 1970s, American firms pursued technological innovation in a variety of ways. Frequently, American industrialists chose innovation as a strategy not as a result of inexorable scientific and technological change but as a response to challenges in the marketplace and from government policy. In some cases, they relied on scientists, but American companies also looked to inventors to develop new products and processes. And in many cases, since research and innovation can disrupt smoothly running operations, American business leaders struggled with where to locate innovation—should they bring research inside the firm or keep it at arm's length? How does one connect the practical, immediate needs of business with the long-term, free-flowing nature of scientific research? By using a selection of historical examples, I want to argue that, while the linear model has preoccupied policymakers, American corporations have pursued—and still continue to pursue—a rich variety of strategies and practices with regard to innovation. In other words, there is no linear history of the linear model of R&D.

PUTTING R&D IN BROAD HISTORICAL PERSPECTIVE

Although this paper focuses on America from the 1870s to the 1970s, I would like to begin by thinking for a moment about R&D across an even longer historical perspective. At its essence, R&D is about pursuing technological change in order to gain an advantage—it's about power. On one level, then, you could argue that R&D dates back to at least the first emperor of China, Shihuangdi (259–210 BCE), who used a variety of technologies—a system of roads, the Great Wall, standardized weights and measures—to consolidate his power.[4] Equally, you could talk about how temples in Alexandria during the Hellenistic era employed inventors such as the Greek Heron to develop mechanical gadgets (such as automatic doors and talking statues) to attract new worshippers.[5]

But as noted in the introduction, R&D is about harnessing technological change in pursuit of economic growth—that a strong economy translates into political power. For this idea, we need to look to Renaissance Italy in the fourteenth century. As Italian city-states found themselves in competition and unable to gain an advantage over each other, several turned to technology. While some cities improved their military technology (in terms of new weapons and fortifications), others sought innovation in manufacture (new products and processes). Offering to provide new technology, artists such as Leonardo da Vinci and Francesco di Giorgio created the role of the inventor. Like artists, inventors claimed that their ability to create new technology was based on personal knowledge and a flash of inspiration—the "Eureka!" moment. Using their artistic training, Renaissance inventors often sketched ideas for new machines. Their ideas could be fanciful, ranging from an undersea diver sketched by Jacopo Mariano ("il Taccola") to Leonardo's flying machines. They also skillfully combined components (such as the wedge, screw, lever, pulley, and gears) to create clocks, sawmills, or a weight-driven spit for turning roasting meat. Leonardo and his Renaissance contemporaries established the ideas that nature could be improved by the human mind and that technology should contribute to progress of society and the state.

From 1500 to 1800, technological change became increasingly important as European states encouraged exploration, military conquest, trade, and manufacturing. These activities depended on better ships, instruments, and weapons, which in turn stimulated the development of new sources of power (coal and the steam engine), better materials (glass and iron), and new ways of organizing labor (factories). To encourage individuals to develop new machines and processes, governments began issuing patents that gave inventors exclusive ownership of their creations. The first patents were awarded by the Republic of Florence in 1421, and the first British patent law was passed in 1623.

To promote technology, different European states pursued different strategies. In France, the strategy was to establish strong national institutions. To consolidate his power, Louis XIV sponsored royal industries in textiles and porcelain, as well as a nationwide system of roads and canals. To design this transportation network, the French established in 1675 a special organization of engineers, the *Corps des ingenieurs du Genie militaire*, and the first engineering school, the *Ecole des Ponts et Chausées*, in 1747. Through these institutions, the French produced talented engineers, but they did not generate an industrial revolution.

In contrast, the British saw invention as the prerogative of the individual, who should be permitted to develop and own new machines. British society would grow wealthy, argued the Scottish economist Adam Smith, if numerous people pursued their individual economic destinies. While much of the British Industrial Revolution was based on countless small changes in the design and manufacture of goods, a few inventors—James Watt, Richard Trevithick, and George Stephenson—concentrated on major developments such as the steam engine and the railway.[6] Well aware of the importance of linking inventions to prevalent beliefs, Watt's business partner, Matthew Boulton, would dramatically tell visitors to their factory that "We sell here, sir, what all the world wants: power."

Just as creative technologists in Britain called themselves inventors, so did ambitious Americans do the same. As early as 1641, American inventors petitioned colonial legislatures for patents. By the end of the American Revolution, British industrialization was well underway, and the Founding Fathers appreciated the importance of stimulating invention. When they framed the Constitution, one of the powers given to the Federal government was to issue patents.

Spurred not only by the patent system but also by firsthand experience of using machines in trade or farming, Americans readily imagined new inventions. As one European visitor remarked, "there is not a working boy of average ability in the New England states ... who has not an idea of some mechanical invention for improvement ... by which, in good time, he hopes to better his position, or rise to fortune and social distinction."[7] Recognizing the importance of agriculture for the new republic, Americans developed machines for harvesting or processing crops. Oliver Evans introduced an automated flour mill in 1790, Eli Whitney patented his cotton gin in 1794, and Cyrus McCormick demonstrated his mechanical reaper in 1831. Meanwhile, Robert Fulton (the steamboat) and Samuel F. B. Morse (the telegraph) contributed the transportation and communications technology needed to sustain an expanding nation.[8] Inventors frequently developed only one or two new devices, which they then put into manufacture themselves or sold to eager entrepreneurs. For instance, a young Philadelphian, Matthias Baldwin, designed a new locomotive in 1830. Finding no one willing to build it for him, he set up his own company that became the leading manufacturer of locomotives in the United States for the next eighty years.[9]

While various individual inventors made their way in the antebellum American economy, their efforts were nonetheless circumscribed. Firms in this period were generally small partnerships and lacked substantial capital. Most industries were marked by sharp price competition, which forced businessmen to avoid the long-term investment needed to improve technology. While businessmen were willing to purchase patents from inventors and put them into use, they generally kept inventors at arm's length, reluctant to employ them or subsidize their development costs. For example, after accidentally discovering the process of vulcanizing rubber in 1838, Charles Goodyear spent an additional five years and $50,000 perfecting and patenting his process. Even though he was able to sell the rights to his patent in both Europe and America, he nonetheless died in 1860 with debts of nearly $200,000. Despite the fact that Americans equated invention with social progress, individual inventors found it hard to negotiate the links between their specific inventions and the needs of business.

WESTERN UNION, MENLO PARK, AND THE ORIGINS OF R&D

In the 1870s, circumstances in the telegraph and electrical industries created a "Golden Age" for inventors and set the stage for the creation of the first R&D facilities.[10] While based on Morse's invention, it was not the heroic origins of the telegraph industry that made it a hotbed of inventors but rather the appearance of the Western Union Telegraph Company. In the 1850s, Morse's telegraph was promoted by numerous small companies, but it soon became clear that the telegraph would only flourish if one system connected cities across America. By absorbing its competitors and building the first transcontinental line, Western Union created a nationwide system in 1867.[11]

But no sooner had Western Union achieved national dominance than it had to fight off critics and rival networks. As a monopoly, Western Union was seen as a threat to American democracy. Critics worried that Western Union controlled the flow of news and stock prices, and that it might use this power to ruin individual businessmen and manipulate the stock market. At the same time, another challenge came from Wall Street. Western Union had expanded by erecting lines along railroads and placing offices in train stations. This meant, however, that as new railroads were built, financiers could create their own telegraph network and attempt to gain control of Western Union. Jay Gould pursued this strategy and eventually captured Western Union in 1881.[12]

In responding to these threats from Wall Street and Washington, Western Union employed various tactics (price competition, political lobbying, and hostile takeovers), but in this turbulent environment, technological innovation came to play a new and important role.[13] To maintain its dominant position, Western Union needed to adopt new inventions that would permit it to operate more efficiently. Likewise, the challengers—financiers and reformers alike—also realized that innovations might be used to gain a foothold in the industry. As the *Telegrapher* observed in 1875:

> improved apparatus has become of vital importance, and, consequently, telegraphic inventors who, for some years past, have been regarded as bores and nuisances, suddenly find themselves in favor, and their claims to notice, recognition and acceptance, listened to with respectful attention. All parties are now desirous of securing the advantages which may be derived from a development of the greater capacity of telegraph lines and apparatus. The fact has become recognized that the party which shall avail itself to these most fully will possess a decided advantage over its competitor or competitors.
>
> That this state of telegraphic affairs affords the opportunity for the inventive talent and genius of the country which has hitherto been wanting, is unquestionable.[14]

By the mid-1870s, the combination of Western Union's dominance and the possibility of a rival network created a unique market for telegraph inventions. There was a strong demand for "blockbuster" inventions that could be used by Western Union or its challengers, and this demand prompted dozens of ambitious men to turn their attention to developing improved devices and entirely new systems. Typical of these inventors was Alexander Graham Bell, who started inventing after reading a newspaper story about Western Union and purchased a patent for a duplex (two-message) telegraph from Joseph Stearns for $25,000 in 1872. While over 400 individuals secured patents for telegraph inventions between 1865 and 1880, the most successful inventors were men such as Thomas Edison and Elisha Gray, who had established themselves as telegraph equipment manufacturers. Western Union quickly came to appreciate Edison's ability to invent new devices and systems, and contracted with Edison to file patents for the company's exclusive use. In turn, Edison used these contracts to leave manufacturing and launch his "invention factory" at Menlo Park, New Jersey, in 1876.

Although Menlo Park is frequently touted as the ancestor of modern R&D labs, its claim to fame turns not on harnessing science but rather because it marks a milestone in linking technological innovation with business strategy. Edison was able to build and operate Menlo Park because Western Union contracted with him to develop several new devices, including a telephone to compete with the new Bell Telephone Company. In this sense, Menlo Park was one of the first facilities to harness technological innovation to corporate strategy. However, Menlo Park was never integrated into the Western Union organization in the way that later labs were. Indeed, Western Union seems to have viewed technological innovation as a risky and expensive proposition, and they chose to minimize their risk by supporting an outside research facility. Thus, Western Union chose to secure innovation through contracts and a strategic alliance with Edison—not unlike semiconductor firms in the 1980s that supported innovation through the Sematech consortium. From Edison's standpoint, not being tied to Western Union was equally desirable, since it permitted him to move into new fields, as he did with electric lighting in 1878. For our non-linear history of R&D, the lesson of Western Union and Menlo Park is that R&D didn't start with science as much as with a realization that technological innovation was a valuable tool for surviving in an environment made turbulent by competition and potential government regulation. Corporate strategy, not science, was the mother of R&D.

For Edison, Menlo Park was an ideal creative environment, and during his seven years there (1876–1883), he turned out a series of spectacular inventions—an improved telephone, the phonograph, and his incandescent lighting system.[15] Skillfully playing up images of the romantic genius for newspaper reporters, Edison gave the American public a highly individualistic myth of technological innovation that perhaps served as an antidote for the realities of the expanding, impersonal organizations (corporations, government agencies, and universities) that were coming to dominate American culture.[16] Edison's success at Menlo Park stimulated other inventors—such as Nikola Tesla and Reginald Fessenden—to set up their own independent laboratories in the 1880s. Even today, American inventors and scientists frequently invoke Menlo Park as the inspiration and model for how they organize their creative efforts.[17]

INVENTORS AND CORPORATE STRATEGY, 1880–1900

As inspiring as they might be, however, Edison and Menlo Park were soon surpassed by other individuals and institutions. Both inventors and businessmen realized that the real challenge in bringing new technology to market lay not with idea generation (research) but with working out the details of manufacturing and marketing (development). While idea generation could take place away from the firm, effective development had to be done inside a firm where one could match the characteristics of a new invention with a company's resources. Consequently, as the high technology of the 1880s—electric lighting—took shape, inventors such as Charles Brush (Brush Electric Light Company) and Oliver Shallenberger (Westinghouse Electrical Manufacturing Company) chose to locate themselves inside new manufacturing companies.

Representative of this new trend of inventors moving into companies was Elihu Thomson.[18] A chemistry teacher from Philadelphia, Thomson was fascinated by the arc lighting systems he saw while visiting Paris in 1878. On his return to the States, he and Edwin J. Houston began developing their own system for lighting factories and shops. Thomson was a successful inventor because of his craft knowledge of electricity; he got ahead because he studied existing devices, carefully constructed his own versions, and then systematically modified these models until he came up with a breakthrough. Although he was trained in chemistry, it was not theory but his hands-on and experimental skills that formed the basis of his technological creativity.

However, Thomson soon realized that while he could invent ingenious devices, he knew little about manufacturing and marketing his creations. Consequently, he allied himself with several different groups of entrepreneurs who provided the funds and expertise needed to commercialize his inventions. After two unsuccessful attempts, Thomson finally found the right set of backers among the shoe manufacturers of Lynn, Massachusetts. Led by Charles A. Coffin, the shoemakers were familiar with marketing since they had developed techniques for selling shoes throughout the United States. Moreover, Coffin was able to secure capital from industrial financiers in nearby Boston and develop new arrangements for extending credit to the newly established utility companies.

Under Coffin's leadership, Thomson was able to concentrate in his Model Room on inventing, and the Thomson-Houston Company grew rapidly. Because many towns and cities in America rushed to create their own local utility companies, there was tremendous demand for electric lighting equipment; in response, Thomson developed new products, including dc and ac incandescent lighting systems, motors, streetcars, and meters. The company built a large factory in Lynn, and by 1891 was employing 2,400 workers. To reach customers throughout the United States and the world, Thomson-Houston established sales offices in major cities and had a large force of salesmen. To help new utilities set up their systems, Thomson-Houston had a construction subsidiary as well as a staff of engineers at the Lynn plant. These many facets of the electrical manufacturing business meant that the Thomson-Houston Company soon came to have a complex management structure, and by 1891, it was capitalized at $10.5 million. For Thomson, the rapid growth of the firm meant that there was a steady demand for his talents as an inventor; new products were needed to reach new markets and compete effectively with the rival companies established by Edison and George Westinghouse. Taking advantage of its efficient organization and Thomson's inventions, Thomson-Houston absorbed most of its smaller competitors in the late 1880s, and in 1892 merged with Edison General Electric to form the General Electric Company (GE).

Thomson demonstrated to Coffin and the other managers at GE that not only could inventors work within the firm but that new products were essential for rapid corporate growth. By the early 1890s, the leaders of GE had come to realize that the size of their firm (in terms of money invested, plant capacity, and organizational complexity) was such that they could no longer hold inventors at arm's length, as Western Union had chosen to do with Edison at Menlo Park. In order to protect their huge investment, product innovation had to be conducted inside the firm. With Thomson as an employee, GE had product innovation inside the firm, and it supplemented Thomson's abilities by hiring more creative inventors and engineers, including Charles Steinmetz.

In terms of our non-linear history, then, the story of Thomson adds several points. First, his story further underlines the idea that firms pursued technological innovation out of strategic considerations—that as firms Thomson-Houston or GE grew larger, they wanted to protect their investment, and in response, GE brought inventors inside the firm. Second, we see with Thomson that innovation in the electrical industry could be performed not just by scientists but by inventors. New products come not only from science but also from craft knowledge.

THE LIMITS OF INVENTION IN THE GIANT FIRM

While the size of the firm prompted GE's managers to support product innovation, size nonetheless interfered with the process of developing new products. As the company became larger, with more factories, departments, committees, and employees, it became increasingly difficult for creative individuals like Thomson or Steinmetz to coordinate the resources they needed to develop new inventions.

This problem is illustrated by Thomson's experience with developing a high-efficiency steam engine and automobile in the mid-1890s. During this period, the U.S. economy was experiencing a severe depression, and GE's primary customers, utility companies, were unable to purchase new equipment. In response, GE developed more efficient generators and lamps, which permitted utilities to make money by lowering operating expenses. GE sought to improve not only its generators but also the engines used to drive them, and the firm asked Thomson to develop a simple engine that could be used in small central stations and isolated plants. Because orders were down for big generators, GE officials hoped that the manufacture of engines might utilize the idle capacity of their large factory in Schenectady, New York.

Thomson quickly realized that an automobile would be an excellent way to test a small engine. For an engine to be successful in an automobile, it would have to be lightweight, simple, and easy to operate. If he could produce an automobile engine with those characteristics, Thomson figured that the same engine would be excellent for powering generators in small stations lacking trained attendants. With the growing popularity of the bicycle as a form of individual transportation, it seemed highly desirable to create a self-propelled vehicle. Along with Henry Ford, Hiram Maxim, and others who took up the challenge of developing a practical automobile, Thomson and Coffin sensed that the success of the bicycle indicated a huge market for a horseless carriage.

After considering electric motors and internal combustion engines, Thomson chose to focus on a steam-powered vehicle. For that vehicle, Thomson designed his "uniflow" engine, which achieved improved thermal efficiency by exhausting cool steam at the end of the stroke through a special set of exhaust ports. By August 1898, the steam vehicle was operational, and Thomson's assistant, Hermann Lemp, demonstrated the vehicle's practicality by driving twenty-five miles, from Lynn to Newburyport and back. Coffin was sufficiently impressed that he encouraged Thomson and Lemp to begin planning for production. In May 1899, they began work on a new, lighter design that was to be "complete and perfect in all parts; in other words to reduce the carriage to a standard article, as if we were building an arc lamp or dynamo for reproduction."[19]

GE, however, chose not to put the Thomson-Lemp steam automobile into production. After consulting with Thomson, Coffin and the company's patent attorneys concluded that they would not be able to secure adequate patent coverage. Although Thomson and Lemp had filed patent applications for details of the vehicle and its engine, it became clear that the company would not be able to assemble a group of patents that would prevent other firms from entering the automobile field. Full-fledged production of vehicles would require a substantial investment by the company, and Coffin believed that it was too risky to make that investment if the company could not control the field. At the same time, GE's core business had begun to recover. The company was receiving new orders from utility companies for equipment, thus eliminating the need for a new product to employ the underutilized plants.

In the course of the automobile project, Thomson grew frustrated with how the company handled new product development. To build and test his engines and automobiles, Thomson had to have different parts made by workers in both the Lynn and Schenectady factories, and then assembled at his laboratory in Lynn. He also had to coordinate with several different groups within the company, such as the manufacturing committee and the patent department, and these groups did not always cooperate. By 1899, Thomson concluded that what was needed was an organizational change, and he wrote to Coffin

> that it has grown upon me strongly within the last four or five months that what is needed is a department at the Works especially for the development of this kind of machinery [i.e., engines]. We should have men and machinery

wholly devoted to work in this field—together with the automobile field—and they should be separated out as it were in a building or department by themselves. As it is, the work is scattered and partly done in one place and partly in another, and it is almost impossible to force it along at the rate required. I find it extremely difficult with the work scattered as it is, to impress upon the men the necessity of saving time or to get a proper appreciation of the value of time in the development of new work. Things move at an exasperatingly slow rate, and the only cause for it that I can discover is the lack of concentration in one place of draftsmen, men and tools.[20]

What Thomson wanted was a department isolated from manufacturing operations, staffed by specialists, and equipped with the necessary machine tools. Although he did not suggest that scientists be hired, what is important is that he wanted a research department as a way of coordinating resources and expediting the innovation process. He had clearly demonstrated that new products could be developed at GE, but because it was such a large and complex organization, he was not able to control the innovation process and deliver new products in a timely fashion. If the firm was to succeed in using new products to gain a competitive advantage, Thomson realized that it would need a new institution suited to the scale of the firm: the R&D department or industrial laboratory. For our non-linear history, Thomson's steam automobile illustrates that the coming of R&D was as much about creating a particular kind of "space" within the firm for product innovation as much as it was about bringing science into the corporation.

GE AND THE FIRST R&D LABORATORY

Thomson discussed his concerns about the organizational arrangements for new product development with the company's other major innovator, Charles Steinmetz.[21] Working first in the calculating department at the Schenectady plant, and then in a laboratory at his boardinghouse, Steinmetz had applied his mathematical skills to improving the efficiency of ac generators, transformers, and motors. By the late 1890s, Steinmetz had become worried that GE's carbon-filament lamp was about to be overtaken by several new and more efficient lighting devices: the Welsbach gas mantle, the Hewitt mercury-vapor lamp, and the Nernst metallic-filament lamp. Aware that those devices had been invented by men familiar with electrochemistry, Steinmetz proposed in July 1897 that the company establish a chemical laboratory where those devices could be investigated. Although his first proposal was ignored by GE officials, Steinmetz repeated his request in 1899, and he enlisted the support of the vice president for engineering, Edwin Wilbur Rice, and the chief patent attorney, Albert G. Davis.

In September 1900, Steinmetz, Thomson, Rice, and Davis succeeded in convincing the company that a research laboratory should be established to investigate and develop new products. To head the new laboratory, the company hired Willis R. Whitney, a professor at the Massachusetts Institute of Technology who had earned his Ph.D. in chemistry at the University of Liepzig under Wilhelm Ostwald. Whitney's mission was to develop immediately a metallic-filament lamp, and he was given a laboratory at the Schenectady plant in 1901 and an annual budget of $15,830. Whitney gradually built up his staff to forty-five scientists and technicians by 1907.

GE created this new entity, the research laboratory, because it was confronted by an immediate competitive threat. If it did not acquire a new high-efficiency incandescent lamp, it was likely to lose a significant portion of the lamp market to Westinghouse and to European lamp manufacturers. To protect its substantial investment in technology, capital, plant, and a skilled workforce, GE had to respond to this threat.

But, like other firms, GE had two choices in how it could respond. Just as Westinghouse had bought the patents for the Nernst lamp, GE could have purchased patents from outside inventors. The other choice was to develop a new lamp in house. GE chose the latter alternative because Thomson had demonstrated that innovation could take place within the firm. Through his many inventions, Thomson had shown the potential of new products for capturing new markets and enhancing the firm's position. Hence, another factor contributing to the creation of the research laboratory was that GE had an established tradition of in-house product innovation.

But it was not enough to have a tradition of innovation. A firm must also have a structure that permits the coordination of people and resources necessary for developing innovations. Thomson's recent experience with automobiles revealed that GE's size and structure were impeding product innovation, thus suggesting that a new kind of laboratory was needed. In order to develop competitive products in a timely fashion, it would be necessary to concentrate resources in a single department. Consequently, a final factor leading to the industrial research laboratory was the gap between the tradition of product innovation and the existing organizational arrangements; because all the activities related to innovation could be performed in the new research laboratory, it was hoped that the new lab would fill this gap.

Although they began in 1900 with high expectations, Whitney and his team found it extremely difficult to develop a better lamp, and in 1906, GE was forced to buy the German patents for manufacturing a tungsten-filament lamp. The lab's first success came in 1907, when William Coolidge demonstrated how tungsten could be made ductile and hence shaped by machine into lamp filaments.

Given that Whitney and the GE lab were unable to contribute any immediate results, why did GE support the lab for the first six years? There are several reasons why, once established, the lab survived. First, besides engaging in basic research, Whitney made sure that his chemists provided the company with a range of services. Along with developing new products for different departments, Whitney and his scientists consulted on production problems, tested materials, and designed pilot plants. The lab actually manufactured some specialty items, such as carbon resistance rods and tungsten contacts, which were used by other parts of the company. And Whitney made sure that his staff filed patents that GE used defensively to protect its existing product line and offensively as bargaining chips in negotiating with rivals. By performing multiple tasks for the firm, Whitney secured the funds needed to subsidize fundamental research.

While GE executives valued these services, they also came to value the laboratory as part of a broader strategy of minimizing risk and uncertainty. As the business historian Alfred D. Chandler, Jr. has argued, giant corporations such as GE grew and survived by performing a wide range of tasks relating to production and distribution. To reduce costs and eliminate uncertainty, firms frequently integrated backward toward their sources of raw materials and forward toward the customer. In pursuing this vertical integration, managers generally chose to bring activities inside the firm rather than to depend on outside suppliers; only by having key functions inside the company did they feel it was possible to minimize risk and protect their large organizations. Given this general strategy, it is not surprising that some managers of technology-oriented firms brought one of the key activities, product innovation, inside the firm. By generating its own new products and patents, a firm ensured a regular supply of this input that it could direct toward increasing productivity and efficiency. Unlike Western Union in the 1870s, which was comfortable in contracting with Edison, GE in the 1900s felt that it could only protect itself by fully integrating innovation into its corporate structure. In this sense, the creation of the industrial research laboratory was part of the early twentieth-century trend in the American economy toward minimizing risk by bringing key functions inside the firm.

But why did big firms like GE invest in a *scientific* laboratory? Why not hire more talented inventors like Thomson and Steinmetz? Here the answer is both economic and cultural. From an economic standpoint, one difference between the 1870s and the 1900s was a change in the supply of scientific manpower. In the 1870s, only a handful of American universities offered advanced research degrees in the sciences, and, like Whitney, those few Americans wishing to become research scientists went to Germany to study physics or chemistry. Yet by 1900, American universities had undergone a profound expansion, particularly in scientific research. Thanks to private philanthropy and the Federal land grants to state colleges, American universities now trained hundreds of Ph.D.-level scientists each year. In fact, George Wise has suggested that the supply of scientists probably exceeded the demand for science professors, and this situation led some scientists around the turn of the century to seek careers in industry.[22] Hence, given the growing supply of scientists, it made sense for managers at GE and other large companies to hire scientists, and not inventors, for product innovation.

But there were also cultural reasons for choosing scientists over inventors. Inventors generally explain and legitimate themselves by claiming that they possess unique personal knowledge (genius) and skills. The basis of their expertise is personal and idiosyncratic. If inventors actually invent in a Eureka! moment, then their work is fundamentally discontinuous and unpredictable. Who knows when the muses will speak? Given this rhetorical stance, inventors were not especially appealing to managers trying to minimize uncertainty and protect companies capitalized for tens of millions of dollars.[23] Yes, a genius like Steinmetz can do great work, but should one bet the company on him?

Instead, along with other attempts to rationalize their organizations, corporate leaders turned to scientists who promised to produce new technology in an efficient and predictable manner. Central to the rhetorical stance of the new industrial scientists of the twentieth century were promises of predictability and continuity. A central characteristic of science was its claim to be able to predict the behavior of natural systems; if this was generally true of science, then the process of applying science to industrial problems should be predictable as well. Moreover, by taking a team approach to solving problems by breaking down complex problems into a series of routine experiments, scientists promised managers that they would get results sooner or later. By promising to be predictable and continuous, industrial scientists spoke a language that made sense to managers who were struggling to protect big firms in the face of uncertainty. The symbolic value of the industrial scientist was perhaps best summed up by Carl Duisberg, director of the research laboratory at the German chemical company, Bayer, who described that his scientists created new dyes so routinely and predictably that "Nowhere any trace of a flash of genius."[24]

The early years of the first R&D lab at GE, then, offer several lessons about the evolution of R&D. Perhaps the most obvious is that while GE hired chemists to develop new products, the scientists failed miserably in the first few years; science did not automatically yield new technology. Given that the lab was not successful, the interesting question then becomes why GE continued to support it. There are several answers to this question; while Whitney found a variety of ways to make science useful to the company, the most important answer is that science-based innovation fit the worldview of the GE managers. Intent on reducing risk and protecting their organization, GE executives found it easier to support the teamwork and routine experiments of scientists than the seemingly irrational activities of inventors. Science, I would argue, came into business not because it was the only way to advance technology, but because it fit the beliefs and values of corporate leaders. It was the cultural function of science, not its economic function, that brought science into the corporation.

CORNING, PYREX, AND THE DISRUPTIVE POTENTIAL OF R&D

Like GE, several other large firms, including Eastman Kodak, AT&T, and DuPont founded R&D labs in the years prior to World War I. In most cases, these companies were seeking to protect their position in the marketplace, and they hoped that scientists could use their expertise to quickly develop new products. For similar reasons, the Corning Glass Works established a laboratory in 1908, but its early experience reveals another facet of our non-linear story: while R&D can improve products, it also can be a disruptive force and create new problems for the company.

Corning dates back to 1851, when Amory Houghton, Sr. organized the Union Glass Works in Somerville, Massachusetts. Hoping to secure fresh capital from local merchants and take advantage of nearby coal deposits, Houghton moved his glass company to Corning, New York, in 1868. There Houghton initially manufactured a range of glass products, including jars, dishes, and lamp chimneys, only to find that the local coal was ill suited to glass-making. This made it difficult for the Corning Glass Works to keep up with the larger glass factories in West Virginia and Ohio, and by 1870, the company was bankrupt. In response, the local creditors reorganized the Corning Glass Works and placed it under the control of Amory Houghton, Jr. and his brother Charles.[25]

Eschewing mass-market products such as bottles, the next two generations of the Houghton family focused instead on specialty glass for industry. From the 1880s to the 1900s, Corning specialized in manufacturing two products—the glass envelopes for incandescent lamps and the lenses and lantern globes used in railroad signal systems. Because railroad lantern globes often cracked as a result of sudden temperature changes (freezing in the winter to hot weather in the summer), Corning was interested in glass that could resist breaking under extreme temperature changes. Knowing that German scientists had discovered how to make glass stronger by using boric oxide, the Houghtons decided in 1908 to create a research department and hired a Ph.D. chemist, Eugene Sullivan, to head up this department. Working with a small team of scientists, Sullivan perfected a heat-resistant lead borosilicate glass, which Corning began marketing in 1909 under the name Nonex.[26]

Nonex lantern globes were immediately popular with Corning's railroad customers. Between 1905 and 1910, Corning increased its share of the railroad glass market from 57 percent to 69 percent. But Nonex's durability had a significant downside—fewer broken lanterns meant less repeat business. From 1906 to 1909, Corning shipped 43,951 dozen clear globes annually to seven major railroads. In contrast, between 1910 and 1913, these railroads purchased 68 percent fewer globes. With approximately one-third of the company's annual revenues tied to railroad glass in 1910, the loss of repeat business was a troubling development.[27] Rather than improving Corning's market position, R&D instead was undercutting Corning's railroad lantern business. Here we see that science-based products do not translate automatically into profits and jobs, as the linear model would suggest.

Corning responded to the collapse of the railroad lantern market by doing more research and by seeking new markets. First, researchers in the R&D lab sought to modify the borosilicate formulation so that it could be used to make new products. By eliminating the lead from its borosilicate glass, Corning scientists developed a new formulation—Pyrex—that could be used to make kitchen bakeware. At the same time, due to the British embargo on German goods coming into America during World War I, Corning used Pyrex to introduce a new line of laboratory glassware that could substitute for German imports.[28] In 1916, Corning sold $286,424 worth of Pyrex goods; in 1917, Pyrex sales jumped to $460,353.[29] Using Pyrex to pursue these two new markets allowed Corning to compensate for the contraction of the railroad lantern business and enabled the company to grow during the 1910s and 1920s. However, this growth taught Corning scientists an important lesson—that scientific research does not automatically produce growth and profitability. Rather, scientists and managers needed to work together to convert laboratory discoveries into successful products.

BLENDING THEORETICAL INQUIRY AND COMMERCIAL NEEDS: LANGMUIR AT GE

As we have seen, the GE industrial research lab during its early years fulfilled several needs of the company, but it did not produce a major breakthrough. All this changed in 1909, when Whitney hired a new young chemist, Irving Langmuir, and assigned him to study why light bulbs acquired a coating on the inside over time. Langmuir's work led GE into the world of electronics and permitted Whitney to define the lab's role in terms of helping GE diversify into new markets.

Langmuir received his Ph.D. from Gottingen University in Germany, where he had studied the behavior of incandescent filaments under the direction of Walther Nernst, the inventor of a metallic-filament lamp. To take advantage of Langmuir's background, Whitney initially assigned him to study the basic physical and chemical processes taking place in the ductile tungsten-filament lamps, but in 1911, Whitney specifically asked Langmuir to study the Edison effect and lamp blackening. In 1880, Edison had observed that the insides of his lamps became black with what were apparently particles of carbon. After some experimentation, Edison decided that these particles were being discharged because there was a current flow from the negative side of the hot carbon filament to the neutral or positively charged interior surface of the bulb. Following Edison, Lee De Forest and J. A. Fleming had used this effect to develop the first vacuum tubes for detecting radio waves, but scientists were debating what caused the effect. Determined to resolve the scientific controversy and to make a name for himself in science, Langmuir pounced on this problem. He found that a hot filament discharged a stream of electrons when the electrons encountered "a sort of subatomic traffic jam just outside the filament surface."[30] This traffic jam was called the "space charge effect," and it had been first observed by C. M. Child at Colgate University. By using the space charge effect, Langmuir was able to not only account for the electron discharges he was measuring on the benchtop but could also improve the design of vacuum tubes and incandescent lamps.

While Langmuir thought about his discoveries in terms of scientific papers, Whitney and another GE manager, Laurence A. Hawkins, were quick to see the commercial advantages of his work. Hawkins in particular saw the connection between Langmuir's filament work and vacuum tubes, and he arranged for Langmuir to work with Ernst Alexanderson, GE's leading radio engineer. As a result, GE soon acquired key patents on radio tubes, which it used during and after World War I to secure a strong position in the new field of radio. Meanwhile, Langmuir convinced fellow researcher William Coolidge to apply his findings about the space charge effect to improve x-ray tubes. Coolidge used this knowledge in 1913 to create a tube that reliably generated more powerful rays. Although GE was initially reluctant to produce the new Coolidge tube on a large scale, it proved to be highly popular with doctors and hospitals, and it led GE to diversify and commit substantial resources to developing x-ray equipment.

The development of both vacuum tubes and the Coolidge x-ray tube marked a new phase not only at the GE lab but in American industrial research generally. Prior to 1913, industrial research labs had been established primarily to protect the company's existing market position by securing patents and improving existing product lines. The mission of the GE lab was to defend the company's electric lighting business by improving lamps. Yet through these new radio and x-ray tubes, the lab permitted GE to diversify and move into new markets. As a result, business leaders came to see industrial research as both a defensive and an offensive tool for corporate strategy.

Whitney and other industrial research leaders realized that Langmuir's work marked a new approach to product development. Previously, they had assumed that professional scientists would improve existing products by using a combination of benchtop and experimental skills, and teamwork. Langmuir, however, was one of the first researchers who got ahead by converting a practical problem (blackening lamps) into a scientific problem (why did a hot filament discharge electrons?) and then translating his scientific knowledge (the space charge effect) back into new products (better vacuum tubes). This blend of the practical and the abstract deeply impressed Whitney and other lab directors, one of whom called this approach "pioneering applied research."[31] Langmuir continued to successfully mix practical problems with scientific research, and in 1932, he was awarded the Nobel Prize in physics. But what Whitney and other industrial research leaders realized through Langmuir was the first confirmation of the linear model—that theoretical science could be performed inside a company and used to create significantly better products.

WALLACE CAROTHERS AND THE SHIFT TO FUNDAMENTAL RESEARCH

During the 1920s and 1930s, stimulated by successful work done at GE, Corning, and elsewhere, R&D expanded steadily across American industry. During the Great Depression, many companies looked to R&D in the hope of gaining new markets through new products or greater efficiency through improved production processes. In 1931, 1,600 companies reported laboratories employing 33,000 scientists, engineers, and technicians, but by 1940, more than 2,000 firms reported R&D departments employing 70,000 people.[32]

While many of these labs probably provided routine testing and support services, DuPont made the next move toward fundamental research by hiring the chemist Wallace H. Carothers. Unlike other industrial scientists who were employed by firms to apply their benchtop skills to practical problems, Carothers was hired by DuPont to conduct fundamental research, identical to the theoretical investigations that scientists were pursuing in American universities. While DuPont executives initially assumed that fundamental research would not necessarily result in profitable new products, Carothers' work did yield two new products: neoprene (artificial rubber) and nylon.

DuPont had established its first R&D facility, the Eastern Laboratory, in 1902 to improve the manufacture of high explosives. Like the GE lab, the Eastern Laboratory grew into the Chemical Department before World War I in order that the company might diversify into new areas such as dyestuffs, celluloid, paints, and artificial leather. By 1927, DuPont was employing 850 people and spending $2.2 million on research, but its research director, Charles Stine, believed that the company should take a more radical approach to R&D. Rather than have professional scientists apply the results of pure science to industrial problems, wondered Stine, why not permit scientists to generate new science in the company's lab?

In proposing to the top management of DuPont that it should fund fundamental research, Stine offered four reasons. First, the company would gain prestige and "advertising value" by being able to claim that its scientists were publishing papers. These public relations concerns were not frivolous, since a portion of the American public saw DuPont as a munitions company that had made fabulous profits from the carnage of World War I. Second, Stine argued that interesting scientific research would make it easier to recruit first-rate Ph.D. chemists who might prefer academic careers. Third, he anticipated that results from fundamental research could be useful in bartering with other companies for patents and proprietary information. And only fourth did he suggest that fundamental research might lead to new products. Convinced by Stine's reasons, DuPont gave him $250,000 to spend annually on fundamental research. Stine built a new laboratory, which was soon dubbed "Purity Hall," and he began recruiting chemists from academia.

Among Stine's first hires was an organic chemist from Harvard University, Wallace H. Carothers. Carothers was initially hesistant about joining Dupont because he was worried that his "neurotic spells of diminished capacity" [what is now called depression] might be a handicap in a corporate environment.[33] However, after receiving reassurances from Stine, Carothers joined the company in 1928 as head of a new group investigating long-chain molecules, or polymers. Just as Langmuir had been attracted by the controversy surrounding electron discharges from hot filaments, so Carothers was excited by the controversy surrounding the nature of polymer molecules. While some chemists thought that polymers were held together by the same forces that operated in smaller molecules, others thought that these large molecules involved some other kind of forces. Carothers resolved this controversy by building long-chain molecules, one step at a time, employing well-understood reactions that used acids and alcohols to form esters. In the course of this research, Carothers and his team not only laid the foundation for our modern understanding of polymers, but in 1930 they also discovered two valuable materials: artificial rubber, or neoprene, and a strong manmade fiber that came to be called nylon.

Under Stine, Carothers' principal obligation was to publish papers about his results, but shortly after the discovery of neoprene and the new fiber, Stine was replaced by a new director of research, Elmer K. Bolton. Bolton had made his reputation at DuPont by converting laboratory research on dyestuffs into commercial products "in the shortest time with the minimum expenditure of money."[34] Bolton had little patience with Stine's ideas about fundamental research, and in response to the Great Depression, he reorganized the groups in "Purity Hall." Bolton believed that new products could be developed faster by combining fundamental and applied research in single teams, and in 1933 he asked Carothers to concentrate his group on developing nylon as a commercial fiber. Carothers did so, but at the personal cost of new bouts of depression. In 1937, just a few weeks after the basic patent for nylon had been filed, Carothers committed suicide.

During the 1940s, nylon came to be used in women's stockings, reinforcement cords in automobile tires, rope, parachutes, and a variety of industrial applications. According to David A. Hounshell and John K. Smith, "[n]ylon became far and away the biggest money-maker in the history of the DuPont Company."[35] Based on its commercial success, DuPont invested heavily in R&D in the 1950s in the hope of getting similar winning products. In doing so, the company assumed that fundamental research would lead to revolutionary products, what had come to be called the linear model of R&D.

Yet the story of Carothers and nylon should be read as a cautionary tale; there is nothing automatic about the conversion of fundamental research to successful product, as the linear model implies. Nylon came about only because of several lucky organizational developments. On the one hand, Stine created a positive environment that would attract a talented chemist like Carothers and that would permit him to do creative research. On the other hand, Carothers might never have converted the fiber he discovered in 1930 into the commercial product nylon without pressure from Bolton. Without the right blend of creative freedom and practical considerations, fundamental research in corporate labs will not yield new "nylons."

BELL LABS AND THE DEVELOPMENT OF THE TRANSISTOR

The importance of blending creative freedom and practical considerations in R&D can also be seen in the development of one of the most famous products of industrial research: the transistor. Invented in 1947, the transistor was the first multipurpose semiconductor device, and it led to the establishment of the modern semiconductor electronics industry. Often touted as the product of theoretical solid-state physics, the transistor is better seen as the result of a mix of physics, business, and hands-on skill.[36]

The transistor was invented at Bell Laboratories, the research arm of AT&T. Established in 1925 as a separate subsidiary, Bell Labs combined AT&T's older Research Department with scientists and engineers from Western Electric, AT&T's manufacturing arm. Bell Labs quickly became the largest corporate R&D lab in the United States, and by the late 1940s, it was employing 5,700 people, of whom more than 2,000 were professional scientists and engineers.[37]

Like other corporate labs, Bell Labs pursued several missions, including solving manufacturing and operations problems, securing patents, and conducting research in areas that would affect the future of telecommunications. One of the future issues that worried Bell Labs was the growing size and complexity of telephone exchanges. By the mid-1930s, the Director of Research at Bell Labs, Mervin Kelly, was becoming concerned that as exchanges grew, mechanical relays used as switches would have to be replaced with electronic devices. While vacuum tubes would be faster than the mechanical relays, Kelly was concerned that tubes would burn out too quickly and draw too much power. Would it be possible, wondered Kelly, to develop an entirely new type of electronics?

Kelly's concerns were temporarily set aside during World War II, during which Bell Labs worked on a wide variety of military projects, including radar. Because radar used microwaves, which could not be detected by vacuum tubes, scientists at Bell Labs, MIT, and Purdue investigated semiconductor materials such as germanium and silicon in order to design new detectors using point contacts. In these investigations, Bell Lab scientists learned how to make two kinds of semiconductor materials (n-type and p-type) by deliberately doping germanium or silicon with traces of other elements. To understand these new materials, researchers drew on a new field of solid-state physics, which used theories and discoveries about electrons and atomic structure to understand the nature of all kinds of materials.

As World War II came to an end, Kelly was anxious to capitalize on the semiconductor expertise that Bell Labs had acquired, and so in 1945, he ordered the creation of a solid-state physics sub-department. The mission of this sub-department was to obtain

> new knowledge that can be used in the development of completely new and improved components and apparatus elements of communications systems ... There are great possibilities of producing new and useful properties by finding physical and chemical methods of controlling the arrangement and behavior of the atoms and electrons which compose solids.[38]

To lead this new group, Kelly selected a chemist, Stanley Morgan, and a physicist, William Shockley. Kelly also assigned two more top-notch physicists,

Walter Brattain and John Bardeen, to the team. While Brattain was an experimental physicist who had worked in the vacuum tube department since 1929, Bardeen was a theoretical physicist who had just joined Bell Labs.

Under direction from Shockley, Brattain and Bardeen focused their efforts on finding a semiconductor device that could amplify signals and thus serve as a possible replacement for vacuum tubes. Shockley first had them try to build a field-effect amplifier by creating a junction of p-type and n-type semiconductors. Shockley hypothesized that when a current was placed across the junction, the induced charge carriers would be free to move, increasing the conductivity of the device and, hence, boosting the signal. When these experiments failed, Bardeen suggested that they did not fully understand what was happening to the electrons on the surface of the semiconductor, and he developed a theory to explain what might be happening. To test this theory, they needed to place two point contacts on the surface of a piece of germanium, and Brattain devised a way to do so by drawing on work he had previously done with point-contact rectifiers. Because Bardeen's theory predicted that the contacts had to be only a few microns apart, Brattain fashioned two contacts by wrapping a plastic triangle with gold foil, cutting a small slit at the apex, and filling the gap with wax. In December 1947, Brattain, Bardeen, and Shockley found that this device could amplify signals. After filing patent applications, Bell Labs announced this invention in June 1948, calling the new device a transistor. Transistor research continued at Bell Labs, culminating in Shockley's development of the field-effect transistor in 1951. The following year, Bell Labs began offering seminars on this new technology and licensing other companies to manufacture transistors. For their pioneering research, Brattain, Bardeen, and Shockley shared the 1956 Nobel Prize in Physics.

Though often celebrated as the classic example of linear R&D, the case of the transistor reflects the organizational factors that make R&D a powerful source of innovation in the United States. While solid-state physics was a necessary ingredient, there first had to be a strategic context in which this knowledge could be utilized. Just as Western Union had turned to technological innovation to maintain its dominant position in the telecommunications industry in the 1870s, so AT&T supported technological innovation at Bell Labs in the 1940s to protect its monopoly position. Next, there had to be not only a tradition of product innovation but also "space" in the organization where creative work could be done. Just as Thomson-Houston had permitted Thomson to work freely in his Model Room in the 1880s, so AT&T established Bell Labs as a freestanding organization in the 1920s. In this way, innovators had their own physical and intellectual space, but at the same time their work was tied to the needs of the firm. In this new organizational space—Bell Labs—the research managers pushed ahead by breaking down complex problems into smaller, more routine tasks and organizing teams of investigators like that of Brattain, Bardeen, and Shockley. As we have seen, the team approach was pioneered by Edison at Menlo Park and perfected by Whitney and Stine at GE and DuPont in the 1910s and 1920s. Of course, the Bell group drew on what they knew, solid-state physics, to develop the transistor, but it is interesting to note that they moved back and forth from the practical goal (getting an amplifier) to a scientific controversy (what was happening on the semiconductor surface) to a new theory (Bardeen's model of surface charges) and finally to a new device (the transistor). In this zigzag course, the Bell team behaved much as their predecessors, Langmuir and Carothers, had done. Like Langmuir and Carothers, the Bell group did not do fundamental research but instead pursued a research program that blended creative freedom with practical needs. In terms of the linear model, the point here is that organizational factors—the strategic context, the position of Bell Labs in the company, the identification of a commercial goal, and the teamwork— were all crucial to the development of the transistor. And when we focus too much on the role of theoretical science in a story like the transistor, we are playing down (and even ignoring) these organizational factors.

R&D IN THE COLD WAR

Emboldened by the success of nylon and the transistor, many American firms invested heavily in R&D in the 1950s. By 1955, total R&D expenditures had risen to $6.1 billion.[39] Like their predecessors at GE, managers during this period continued to see investment in science as part of a risk-averse strategy. As John K. Smith has observed, "if basic science was the seed of new technology, then the entire innovation process could be contained within the firm; reliance on unpredictable outside sources of technology was no longer necessary."[40] American firms found further incentives as a result of antitrust litigation. During the New Deal, the Federal government had attacked a number of firms, including AT&T, DuPont, RCA, and Corning. In this political environment, American firms often found it preferable to develop new products in-house and avoid acquiring new technology through corporate acquisitions or cooperative arrangements.

Not only did the Federal government shape corporate R&D through antitrust litigation, it also influenced it through defense spending. During the Cold War, the Federal government spent hundreds of millions of dollars not only on complex new weapons systems but also on developing new manufacturing techniques. Significantly, the Federal government often financed the risky and expensive commercialization phases of a new technology; for instance, Bell Labs was only able to perfect manufacturing techniques for silicon transistors with support from the Pentagon. One commentator estimates that during the 1950s and 1960s, one-half to two-thirds of the funds for R&D came from Washington.[41]

In investing in R&D, American companies employed thousands of Ph.D. scientists and built elaborate research campuses. At these new facilities, scientists were granted a large degree of autonomy, in the belief that such freedom had been the crucial ingredient in the development of nylon and the transistor. And yet despite ample funds, new facilities, and unprecedented freedom, scientists at the major corporate labs came up with few major break-throughs from the 1950s to the 1980s. Representative of the woes experienced by many American firms was RCA's struggles in the 1970s with the videodisc. Building on its previous successes in consumer electronics with radio, black-and-white television, and color television, RCA poured tens of millions of dollars into this project, only to discover that American consumers preferred the videotape systems developed by the Japanese firms Sony and JVC.[42] Instead, many of the major blockbuster innovations in this period—such as the integrated chip, the personal computer, the laser, and the birth control pill—were introduced by individual inventors and small start-up firms.[43] But before celebrating small start-up firms in the electronics industry as triumphs of the free market, historian John K. Smith reminds us that these small firms were often highly dependent on military funding and on Bell Labs for providing information and personnel.[44]

CORNING AND CELCOR IN THE 1970s

Not only did the Federal government shape the direction of R&D through antitrust litigation and military spending, it shaped R&D through regulation. While it is easy to assume that regulation discourages innovation by business, it can also create opportunities for innovation. Such was the case with Corning and the development for Celcor for use in automobile catalytic converters.

The story of the Corning catalytic converter began in 1970, with a meeting between a Corning executive, Tom MacAvoy, and Ed Cole, the president of General Motors. As mentioned in the introduction to this chapter, Corning had developed a chemically strengthened glass, Chemcor, that MacAvoy hoped to sell to GM for use in windshields. Cole had declined the new glass, arguing that GM had no desire to replace its existing windshields with the more expensive Chemcor version.

But as MacAvoy got up to leave, Cole asked him a question. Cole knew that Corning was designing a ceramic heat exchanger, Cercor, to work with experimental gasoline turbine engines; could Corning use the ceramic substrate in Cercor to make a catalytic converter for automobile engines? Congress was about to enact antipollution legislation (the 1970 Clean Air Act) that would require reduced emissions in all cars starting with the 1975 model year. Yet at that time, no one knew how to make a material that could withstand the intense chemical reactions needed to remove pollutants from automobile exhaust. Perhaps Corning could make the antipollution device that Detroit desperately needed to satisfy the impending legislation.[45]

The challenge to develop a ceramic substrate for catalytic converters came at a time when Corning was going through a series of wrenching changes. Since the 1950s, Corning had been a major supplier of picture tubes to American television manufacturers. By the late '60s, however, American TV manufacturers were facing stiff competition from Japanese electronics firms that were offering cheaper and better sets to American consumers. Hoping to stave off the Japanese competition, RCA (one of Corning's biggest customers) decided to cut costs by integrating backwards and manufacturing its own tubes. In 1968, RCA decided to make its own nineteen-inch tubes, which was the most profitable tube for Corning to manufacture and sell to RCA. Thus, as the '60s came to a close, Corning's "cash cow"—television—was dying.

As MacAvoy and the top management studied the situation in 1970, they saw the catalytic converter as both an opportunity and a risk. On the one hand, developing a catalytic converter was a great opportunity because Federal law mandated that every new car in America would need a converter. If Corning succeeded, there would be a huge demand for this new product. On the other hand, pursuing a substrate for a catalytic converter was an extremely risky decision because

- no one knew what kind of material might work or how it would be fabricated;
- competition would be fierce since several other companies were racing to produce their own converters;
- the product had to be delivered in a very short time (three years); and
- the automakers were hoping that within five years the antipollution legislation would be rolled back and converters would be obsolete.

Yet despite these risks, Corning's managers decided to take the risk and devoted significant resources to the project. Over the next four years, over 50 percent of the Research, Development, & Engineering (RD&E) staff worked on some aspect of the substrate project.

The main difficulty lay in finding a substrate that could stand up to the harsh chemical reactions taking place in the converter. To reduce exhaust emissions, the automotive industry was concentrating on developing several different catalytic converters. In these devices, hot exhaust from the engine was exposed to a catalyst that started a chemical reaction; taking advantage of the high temperature of the exhaust, a catalyst such as platinum would cause a reaction, which would convert the emissions into water and nonharmful gases. Initially, automakers had tried using a metal substrate to hold the catalyst, only to find that leaded gasoline destroyed both the catalyst and the substrate. To partly solve this problem, GM convinced Exxon to start refining unleaded gas, and it decided to produce automobiles that would run on this fuel.[46] While it was well known among materials researchers that ceramics could withstand high temperatures and corrosive environments, could Corning researchers quickly find a ceramic material that would survive inside the catalytic converter?

To find the right ceramic, Corning managers mobilized hundreds of scientists and engineers and deployed them in dozens of short-term research teams.

These teams pursued a combination of divergent and convergent strategies: let a thousand flowers bloom, and then select the most promising for full development. Within the lab, Corning scientists began by brainstorming and generating a variety of approaches. According to project manager Dave Duke,

> We had a program then of trying about four/five/six different approaches to making this ceramic product. That was kind of Bill Armistead's approach: let's try various things and see which ones work. So we started some things that didn't make sense and some that did seem to make sense ... we started making these things using the same kind of technology that we had to make these gas turbine regenerators, these big [Cercor] regenerators, by stacking up layers of a sort of a crimped paper with ceramic on it and burning it out. We tried wrapping that paper up into little circles. We tried extruding bundles and tubes of glass which we called packet, and then firing those to become a glass ceramic. We tried drizzling glass into funny shapes to get a high surface area. There were just a lot of different ideas. We were working both on the process and on the materials ... We were firing and trying all kinds of different things.

In less than a year, one of the research groups hit upon a promising material. In early 1971, Irv Lachman and Ron Lewis developed a particular ceramic, cordierite, out of magnesia, alumina, and silica. Cordierite, they found, withstood the high temperatures and harsh chemical reactions in the converter.[47] Moreover, cordierite looked highly promising from a financial standpoint; the raw materials for making cordierite cost fifty to sixty cents per pound, and the finished product could be sold for about four dollars per pound.

But no one knew how to fabricate cordierite and give it the necessary form. For a catalytic converter to work efficiently, the exhaust gases must have maximum exposure to the catalyst that is on the surface of the substrate. To achieve maximum exposure, the substrate has to have an enormous surface area, creating long passages for the exhaust to travel through. To get this huge surface area in a small space, the substrate must be very thin. Initially, many people at Corning thought they could borrow ideas from the honeycomb-structured substrate used in the Cercor converter. However, while the honeycomb structure was acceptable, the methods of fabricating the Cercor structure did not work with cordierite. In the Cercor converter, the ceramic was formed around a cardboard form, and when the ceramic was fired, the cardboard burned away. Since the substrate made from cordierite had to be extremely thin, it could not be shaped using this technique from Cercor.

The solution to this new problem was developed by another ceramics researcher, Rod Bagley, who came up with the idea of extrusion. While often used with plastics, no one had ever tried to extrude a ceramic and form very thin walls separated from each other by only a few hundredths of an inch. Bagley had to develop an extrusion die capable of withstanding tremendous pressures. Using this new extrusion die, Bagley was able to create the necessary thin but intricate honeycomb shape needed for the catalytic converter. Once the management team saw that cordierite could be extruded, they decided to shift from a divergent to a convergent research strategy. As Duke recalled,

> [A]s soon as we saw the first little samples, that they could be extruded—we all looked at it and said, "Wow. This is the way to go." Instead of debating it and talking about it for a month or two like most people would do, we sat down and Bill Armistead said, "This is clearly the way to go. Let's take all the people off all the other ones."[48]

Armed with a new material—cordierite—and a new fabrication technique—extrusion—Corning turned to manufacturing its new substrate for catalytic converters. Roughly the size of a coffee can, the converter contained a honeycomb cordierite structure with 200 rectangular cells per square inch that provided the surface area of a football field. Throughout the extrusion process, the walls in the structure had to maintain their exact dimensions so that the platinum catalyst could be applied evenly and economically, so that the exhaust gases would flow smoothly through without meeting obstructions that could cause back pressure.[49] Moreover, because automakers needed to put a catalytic converter on every new car manufactured in 1975, they needed millions of substrates per year, and Corning engineers had to figure out how to produce these quantities quickly, reliably, and cheaply. Working furiously, Corning broke ground on a new, special plant in January 1973 and shipped its first units in April 1974. Named Celcor, this new substrate material was first installed in cars for the model year 1975.[50] Celcor proved to be highly profitable, and by 1994, it had generated $1 billion in sales for the company. Since then, catalytic converters using Celcor have helped to reduce the level of air pollution in the United States and around the world.

The story of Celcor offers several final lessons in terms of a non-linear history of R&D. First and foremost, we see again how R&D is not particularly shaped by the supply of scientific knowledge but rather by changes in the larger context—here by opportunities created by new environmental regulations. Next, we should note that while the search for a catalytic converter substrate was risky, the need was well defined—a phenomenon that we have noted before with GE and the incandescent lamp or AT&T and the transistor. Within the project itself, there wasn't a whole lot of theorizing going on; the project involved a long series of trial-and-error experiments, much like the approach taken by Edison at Menlo Park. Equally, the problem of how to form the new material was based on borrowing extrusion from the plastics industry and figuring out how to extrude the material using high pressures and to a high level of precision. Hence, even in the 1970s, in a company well known for its scientists, we see little that looks like the linear model of fundamental research to commercial product.

CONCLUSION

At the start of this chapter, I hypothesized that there is no linear history to the linear model of R&D, and I proposed to use a series of cases from American business over the last 100 years to investigate this proposition. So what have we found out?

Cracks in the Linear Model

I would suggest, first, that the stories told here don't provide much evidence to support the set of relationships that comprises the linear model. Let's begin with the relationship between scientific theory and new products. In a few cases—such as Carothers and nylon or Bell Labs and the transistor—science did provide the theory that led to new products. However, in other cases, new technology did not depend on theory at all. Edison and Thomson revolutionized the electrical industry without resorting to scientific theory and instead relied on careful observation and craft knowledge. Lest one think that craft knowledge only really belongs to the 1870s, I would point out that the development of Celcor in the 1970s turned not on fundamental research as much as it did on Edisonian trial and error.

If it isn't scientific theory that companies use to create new products, then why do they employ so many scientists? How do scientists add value to the R&D enterprise? The cases here reveal that scientists contribute new benchtop skills, an experimental methodology, and teamwork. In several cases mentioned here, companies hired scientists because they had acquired in the course of their Ph.D. research the ability to manipulate new phenomena. For example, Whitney hired Langmuir because he had learned from Nernst how to conduct experiments with incandescent filaments. Equally, Bardeen played a crucial role in the invention of the transistor by being able to rig up a device to test his colleagues' theory. Although I can only cite anecdotal evidence from my experience of teaching in a research-driven engineering school, I would further hypothesize that universities have come to play such a strong role in new industries such as biotech and nanotechnology not because professors impart theory to the students but because Ph.D. students develop the requisite benchtop skills in university laboratories.

Scientific training also gives researchers other attributes that are valuable for R&D—the experimental method and teamwork. Scientists are good at breaking problems down into a series of tests or experiments that can be conducted by a whole team of investigators. As we have noted, risk-sensitive corporate managers readily appreciate that these attributes help make R&D at least seem to be more routine and predictable. Hence, we should not overlook the cultural efficacy of science in the corporate setting—that science came into the corporation because it complemented the worldview of the managers running large-scale capital-intensive enterprises.

Let's turn to the middle relationship in the linear model—that business simply applies science to create new products. Embedded in this relationship is the assumption that it should be a relatively smooth and even automatic process to move ideas or results from the laboratory to the marketplace. But as we have seen, firms often struggle in managing this transition, whether it be Corning and Chemcor or GE and the steam automobile. In the case of DuPont and nylon, we saw that the story depended on getting, by chance, the right mix of managers—while Stine provided the encouragement that Carothers needed for risky cutting-edge research, Bolton provided the bottom-line discipline needed to convert nylon into a commercial product. Likewise, Armistead and Duke managed the Celcor project at Corning so that a large number of researchers stayed focused and produced out a product in time. The point here is that science-based research does not proceed automatically—it requires creative and thoughtful intervention by managers who can integrate information about the research and the marketplace.

Finally, let's not forget the last relationship in the linear model: that new products are supposed to create profits and jobs. Here the relevant example is Nonex at Corning in the 1910s. Based on scientific research, Nonex was a superior product; but rather than improving Corning's position in the railroad lantern market, Nonex actually undermined it. The lesson here is that there is no guarantee that scientific research leads automatically to profits or employment; it again depends on carefully orchestrating research with market demand. Corning survived because it was able to convert Nonex into Pyrex and go after new markets in terms of bakeware and laboratory glassware. To be sure, Nonex is only one example, but it should serve to remind us that fundamental research doesn't automatically generate profits or jobs.

THE SHAPE OF R&D HISTORY

So while the cases reveal flaws in each part of the linear model, what do they say collectively about the history of R&D? Is the trajectory of R&D over the last century linear or non-linear? Does it proceed in any sort of straightforward fashion?

As a first step, we can experiment with plotting the cases discussed here. A few years ago, Donald E. Stokes critiqued the linear model by developing a 2×2 matrix that categorized R&D activities according to motivation (see Figure 3.1).

In his matrix, Stokes plotted three kinds of research and related them to individual exemplars; hence the quadrant for pure basic research is named after the theoretical physicist Niels Bohr, the quadrant for use-inspired research is represented by the chemist Louis Pasteur and the quadrant for pure applied research is named after Thomas Edison.[51] One could argue that a linear history of R&D would consist of moving from pure applied research to use-inspired research to pure basic (Edison to Pasteur to Bohr) (Figure 3.2).

Consideration of Use?

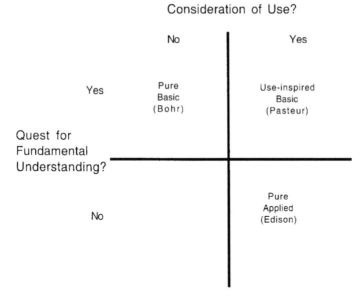

FIGURE 3.1.   Stokes Matrix Model (*Source*: Donald E. Stokes, *Pasteur's Quadrant: Basic Science and Technological Innovation* [Washington, D.C.: Brookings Institution Press, 1997].)

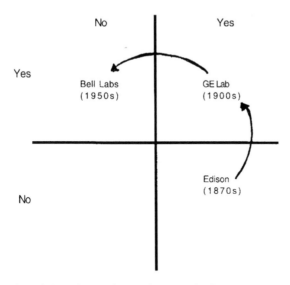

FIGURE 3.2.   What Might a Linear History of R&D Look Like?

In contrast, if we plot the cases discussed here, we get a very circuitous loop (Figure 3.3). Clearly, over time, American companies have tried different strategies and moved in and out of the various quadrants. Naturally, the cases

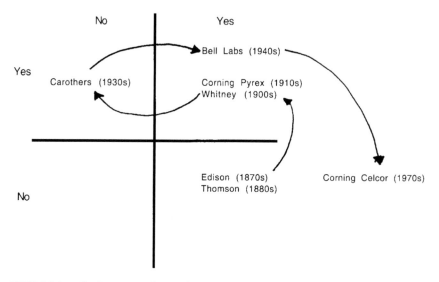

FIGURE 3.3. Plotting Cases Discussed Here

in this chapter are only a small and hardly a random sample, but I suspect plotting an even larger number of cases would not produce an even more circuitous pattern.

So what does this circuitous loop suggest? The first thing it reveals is that the history of R&D is not a supply-side story—that R&D was response to the growth of science in the last 100 years. If it were a supply-side story, then most of our cases would be in the Bohr quadrant. Although chemistry and physics grew enormously in the late 19th and early 20th centuries—in terms of practitioners and volume of knowledge—this growth did not lead inexorably to the creation of corporate R&D laboratories. Corporate mythology and older historical scholarship assumed that the "science-based" electrical and chemical industries could only be developed by the application of science. Once the infant electrical or chemical industries were established, so the argument ran, it was only a matter of time before the industrialists would have to hire scientists to develop better products.

Yet across these cases drawn from the electrical and chemical industry, nowhere have we seen the volume or availability of scientific knowledge force companies to take up scientific research. In direct contrast to this deterministic view, I have suggested here that new product development could be done by either inventors or scientists and that firms chose to hire scientists for a variety of reasons. To be sure, George Wise was right in arguing that companies like GE may very well have hired scientists for product innovation because of the growing supply of scientists. However, managers shifted from inventors to scientists in part because the image and rhetoric of science appealed to managers intent on protecting their organizations by minimizing risk.

Instead, I think that the history of R&D is a demand-side story—that's why most of the cases are in the Pasteur and Edison quadrants. R&D came about, by and large, not because of the growing volume of scientific knowledge but because of the problems faced by large-scale organizations. Anxious to maintain dominant positions in their respective industries or to protect their significant assets, companies turned to technological innovation as a strategy—just look at the reasons why Western Union supported Edison or GE hired Whitney to start its first R&D lab. More often than not, technological innovation has been part of a defensive strategy in the sense that firms hope that innovation will permit them to maintain the status quo. Obviously, the research undertaken by Langmuir on vacuum tubes, Carothers on nylon, and Bell Labs on the transistor all resulted in new products and new industries, but we should not overlook that their respective companies supported this research in the hope of making effective use of their existing capacity. It seems to me that there are far fewer cases in which firms took up innovation offensively, that is, deliberately in order to create new products or new industries—and the one example developed here is Corning and Celcor.

It is important to take in what it may mean to say that the history of R&D is a demand-side story. On the one hand, most people favor greater investment by both the government and companies in R&D since they do want to believe the fundamental idea in the linear model—that more science and technology will give us a stronger economy. As I observed early on, this is an idea that has guided political and technological thinking for centuries, since the Italian Renaissance. And modern R&D is supposed to give us revolutionary new products that improve daily life, shake up industries, and ultimately grow the economy. On the other hand, the historical cases here reveal that firms generally invest in R&D in response to their own organizational needs and in order to maintain the status quo. Yes, firms want new technology, but only if it allows them to maintain or improve their existing market position. Companies are not in the wholesale revolution business. *Plus ça change, c'est la meme* (the more it changes, the more it is the same).

It is vital that we think about the history of R&D not just for historiographic reasons but also for policy. In writing this paper, I came across the National Research Council's 1999 report *Harnessing Science and Technology for America's Economic Future*. In this report, I was pleased to read that experts at the NRC symposium were suspicious of the linear model and noted that "the process of harnessing science and technology for economic growth is complex and not adequately understood."[52] However, even though they acknowledged the problems with the linear model, the economists and policymakers seemed to assume that more money spent on science and technology should automatically produce economic growth; in fact, I thought they were even more vague on how the money should be spent than Vannevar Bush had ever been.

Vague generalizations still guide much of the thinking about the relations between science, technology, and economic growth. In this situation, we historians have a responsibility to help policymakers learn from specific episodes and to frame policies that reflect how science and technology actually get done. To paraphrase the environmental historian Donald Worster, "If historians per se have anything special to add to [economic] analysis, it is the awareness that all generalizations must be rooted in specific times and places—not a small point when there are avid generalizers about."[53] A realistic alternative to the linear model of R&D is something that the world really needs.

NOTES

1. Margaret B. W. Graham and Alec T. Shuldiner, *Corning and the Craft of Innovation*. New York: Oxford University Press, 2001, pp. 260–3.

2. Vannevar Bush, *Science: The Endless Frontier*. Washington, USGPO, 1945, p. 5.

3. Examples of this older approach to the history of R&D included Leonard S. Reich, *The Making of American Industrial Research: Science and Business at GE and Bell, 1876–1926*. New York: Cambridge University Press, 1985; J.D. Bernal, *Science and Industry in the Nineteenth Century*. Bloomington, Ind.: Indiana University Press, 1970.

4. Francesca Bray, "Early China," in W. B. Carlson, ed., *Technology in World History*. New York: Oxford University Press, 2005, vol 2.

5. Peter James and Nick Thorpe, *Ancient Inventions*. London: Michael O'Mara, 1995, pp. 128–139.

6. Maxine Berg, *The Age of Manufactures, 1700–1820: Industry, Innovation, and Work in Britain*, 2nd ed. London: Routledge, 1994.

7. Quote is from George Wallis, an English engineer and manufacturer. See Marvin Fisher, *Workshops in the Wilderness: The European Response to American Industrialization, 1830–1860*. New York: Oxford University Press, 1967, p. 48.

8. Brooke Hindle and Steven Lubar, *Engines of Change: The American Industrial Revolution*. Washington, D.C.: Smithsonian Institution Press, 1986.

9. John K. Brown, *The Baldwin Locomotive Works, 1831–1915*. Baltimore: Johns Hopkins University Press, 1995.

10. Thomas P. Hughes, *American Genesis: A Century of Invention and Technological Enthusiasm, 1870–1970*. New York: Viking-Penguin, 1989, pp. 13–95.

11. Robert Luther Thompson, *Wiring a Continent: The History of the Telegraphy Industry in the United States, 1832–1866*. Princeton: Princeton University Press, 1947.

12. W. Bernard Carlson, "The Telephone as a Political Instrument: Gardiner Hubbard and the Political Construction of the Telephone, 1875–1880," in M. Allen and G. Hecht, eds., *Technologies of Power: Essays in Honor of Thomas Parke Hughes and Agatha Chipley Hughes*. Cambridge, Mass.: MIT Press, 2001, pp. 25–55, and Richard R. John, "The Politics of Innovation," *Daedalus* 127 (Fall 1998): 187–214.

13. Paul Israel, *From Machine Shop to Industrial Laboratory: Telegraphy and the Changing Context of American Invention, 1830–1920*. Baltimore: Johns Hopkins University Press, 1992.

14. "The Progress of the Telegraphic Contest," *The Telegrapher* 11 (Jan. 30, 1875), p. 28.

15. William S. Pretzer, ed. *Working and Inventing: Thomas Edison and the Menlo Park Experience*. Dearborn, Mich.: Henry Ford Museum and Greenfield Village, 1989; Robert Friedel and Paul Israel, *Edison's Electric Light: Biography of an Invention*. New Brunswick, N.J.: Rutgers University Press, 1986.

16. Charles Bazerman, *The Languages of Edison's Light*. Cambridge, Mass.: MIT Press, 1999.

17. Kenneth A. Brown, *Inventors at Work: Interviews with Sixteen Notable American Inventors*. Redmond, Wash.: Tempus, 1988.

18. W. Bernard Carlson, *Innovation as a Social Process: Elihu Thomson and the Rise of General Electric, 1870–1900*. New York: Cambridge University Press, 1991.

19. Thomson to Coffin, 11 May 1899, Letterbook 4/99–7/1900, p. 107, Elihu Thomson Papers, Library of the American Philosophical Society, Philadelphia.

20. Thomson to Coffin, 12 September 1899, Letterbook 4/99–7/1900, pp. 371–4, Thomson Papers.

21. Ronald Kline, *Steinmetz: Engineer and Socialist*. Baltimore: Johns Hopkins University Press, 1992.

22. George Wise, "A New Role for Professional Scientists in Industry: Industrial Research at General Electric, 1900–1916." *Technology and Culture* 21: 408-29 (1980).

23. Louis Galambos, "The American Economy and the Reorganization of the Sources of Knowledge," in A. Oleson and J. Voss, eds., *The Organization of Knowledge in America, 1860–1920*. Baltimore: Johns Hopkins University Press, 1979.

24. Quoted in Henk van den Belt and Arie Rip, "The Nelson-Winter-Dosi Model and Synthetic Dye Chemistry," in W. E. Bijker, T. Pinch, and T.P. Hughes, eds., *The Social Construction of Technological Systems*. Cambridge, Mass.: MIT Press, 1987, pp. 135–58 on p. 155.

25. Davis Dyer and Daniel Gross, *The Generations of Corning: The Life and Times of a Global Corporation*. New York: Oxford University Press, 2001, pp. 24–47.

26. Graham and Shuldiner, pp. 41–46; 54–55.

27. Dyer and Gross, p. 95.

28. Graham and Shuldiner, pp. 55–60.

29. Dyer and Gross, p. 102.

30. George Wise, *Willis R. Whitney, General Electric, and the Origins of U.S. Industrial Research*. New York: Columbia University Press, 1985, p. 175.

31. This term was used by Charles Stine at DuPont in the 1920s. See John Kenly Smith, Jr. and David A. Hounshell, "Wallace H. Carothers and Fundamental Research at Dupont." *Science* 229: 436–42 (August 2,1985), p. 436.

32. Kendall Birr, "Industrial Research Laboratories," in N. Reingold, ed., *The Sciences in the American Context: New Perspectives*. Washington, D.C.: Smithsonian Institution Press, 1979, pp. 193–208, on p. 199.

33. Quoted in David A. Hounshell and John Kenly Smith, Jr., *Science and Corporate Strategy: DuPont R&D, 1902–1980*. New York: Cambridge University Press, 1988, p. 230.

34. Smith and Hounshell, "Carothers and Fundamental Research," pp. 439–440.

35. Hounshell and Smith, *Science and Corporate Strategy*, p. 273.

36. Michael Riordan and Lillian Hoddeson, *Crystal Fire: The Birth of the Information Age*. New York: W.W. Norton, 1997.

37. Ernest Braun and Stuart Macdonald, *Revolution in Miniature: The History and Impact of Semiconductor Electronics,* 2 ed. New York: Cambridge University Press, 1982, p. 33.

38. Quoted in Dirk Hanson, *The New Alchemists: Silicon Valley and the Microelectronics Revolution*. New York: Avon, 1983, p. 74.

39. Birr, p. 202.

40. John Kenly Smith, Jr., "The Scientific Tradition in American Industrial Research." *Technology and Culture* 31:121–31 (1990), p. 128.

41. Birr, p. 202.

42. Margaret B. W. Graham, *RCA and the VideoDisc: The Business of Research*. New York: Cambridge University Press, 1986.

43. David E. Brown, *Inventing Modern America: From the Microwave to the Mouse*. Cambridge, Mass.: MIT Press, 2002.

44. Smith, p. 130.

45. Dyer and Gross, pp. 302–305.

46. Graham and Shuldiner, p. 352.

47. Graham and Shuldiner, p. 353.

48. Quoted in Graham and Shuldiner, p. 354.

49. Dyer and Gross, p. 323.

50. Dyer and Gross, p. 324.

51. Donald E. Stokes, *Pasteur's Quadrant: Basic Science and Technological Innovation*. Washington, D.C.: Brookings Institution Press, 1997.

52. National Research Council, *Harnessing Science and Technology for America's Economic Future*. Washington, D.C.: National Academy Press, 1999, p. 18.

53. Donald Worster, "History as Natural History: An Essay on Theory and Method," *Pacific Historical Review* 53:1–19 (Feb. 1984). I have substituted the word economic where Worster used ecological.

4

Silicon Valley's Next Act: Creativity, Consumers, and Cross-Disciplinary Innovation Move toward Center Stage

KIM WALESH

P erhaps no region in the world has undergone more profound change in fifty years than San Jose/Silicon Valley. Since the 1950s, the area from Palo Alto to San Jose has transformed from an agricultural economy to the world's leading center of technology innovation and entrepreneurship.

Waves of innovation—in defense electronics, integrated circuits, personal computing, the Internet, and networking—powered business start-up and growth and fueled in-migration from across the United States and around the world. Small, disconnected communities grew together into a well-known region of more than 2.3 million people. The region's urban center, San Jose, grew from 95,000 people in 1950 to nearly 950,000 in 2005, when it emerged as the tenth largest city in America. Despite the employment contraction following the 2001 dot-com bust, the region remains the most significant concentration of technology companies and talent in the world.[1]

Yet today, business and civic leaders in Silicon Valley are readying for the next wave of change. Companies, both established and brand-new, are tapping technical talent and courting expanding markets around the world as the global business model advances in its next stage of evolution. As a relatively high-cost region and mature technology center, Silicon Valley is challenged to provide a very high "return on location" for companies that operate here—an environment for innovation, entrepreneurship, and productivity that can remain unparalleled when compared to a growing set of competitors.

Listening to their corporate community, regional leaders are seeing creativity, consumer experience, and cross-disciplinary innovation as central to sustaining Silicon Valley's unique role as the world's leading center for innovation. This awareness is sparking new dialogue and actions to create new "place-based" advantages that can support the next wave of innovation.

THE IDEA ECONOMY VALUES CREATIVITY

In many ways, Silicon Valley has been, for a long time, the ultimate "idea economy"—a place where companies and communities have grown through developing and using new ideas. Since the early days of Hewlett, Packard, the Varian brothers, and Fairchild Semiconductor, the value of technology products invented here has come not from the physical inputs themselves, but from knowledge and intellectual capital that combine and augment basic physical materials (e.g., silicon) in powerful ways.[2]

This idea-based economy values creativity—the ability to generate new ideas, and to link ideas in novel, nonroutine ways. While creativity may be perceived traditionally as the realm of the artist, creativity in a general sense has become fundamental to devising new products, services, and technologies. High-end creative capacities—such as originality, divergent thinking, advanced conceptualization, synthesis, tolerance of ambiguity, remote association, and intellectual curiosity—are essential for local companies' competitive success, whether they are creating an innovative new chip architecture, a new software application, or a new search functionality.

While entrepreneurs and executives in Silicon Valley have always emphasized innovation, company executives and employees today talk increasingly about creativity as key to value creation in this region. In a recent survey of Silicon Valley tech workers, 84 percent said that creativity is important to the success of their business.[3] And the 2006 *Index of Silicon Valley* documents that Silicon Valley has a much stronger concentration of design, engineering, scientific, and business management talent to drive the creation of new ideas, methods, products, services, and business models than do other technology centers. This high-end talent comprises 14 percent of overall Silicon Valley employment, compared with 8–9 percent of the employment base in the next-closest regions of Austin, Seattle, and San Diego. Especially in high-cost regions like Silicon Valley, innovative companies must marry disciplined creativity and high-order value creation with aggressive commercialization.

TECHNOLOGY + DESIGN/CONSUMER EXPERIENCE

More and more, product value stems not just from a product's creative new technical features, but from the product's design and other immaterial qualities that please consumers. Nontechnical elements—design, ease of use, brand, personalization, quality of service, distribution experience, content—are becoming more important ways of creating and sustaining competitive advantage for technology products.

The ability to combine deep technical knowledge with new design skills and consumer orientation will be essential for Silicon Valley companies, and for the region itself, to succeed. As Leslie Bixel, an Adobe executive overseeing innovation in the Advanced Technology Group, explains, "Having the coolest technology is important, but so is delighting the consumer with the entire product experience. This means user-centered design and more focus than in the past on distribution, marketing, and brand."

This newfound importance on design and sensitivity to consumer experience is a departure from Silicon Valley's history as primarily a producer economy. Traditionally, most Silicon Valley companies produced products that were sold to other businesses, and were then used as inputs to final products or for production support (e.g., semiconductors, electronic components, semiconductor equipment). Today, a growing segment of Valley companies is now focused on consumers. Some of these, such as Yahoo!, eBay, and Google, emerged during the Internet boom. Others, "old" by Silicon Valley standards, are energized around new consumer products—Apple with its iPod and iTunes; Hewlett-Packard's emphasis on digital photography and video tools for personalized artistic creativity and content creation; Adobe with software tools for creating, editing, and distributing digital images, audio, and text; Electronic Arts with computer and video games; IDEO's international prominence in product design. Even some producer product companies, such as Intel, are making significant investment in the "soft" technology of consumer branding.

Paralleling the new emphasis on design and consumer experience has been a solid contraction in production-related employment and a rise in software, business and information services, and headquarters functions.

REQUIREMENTS FOR SUCCESS

This new valuation of creativity and the consumer experience emerged in the wake of the dot-com collapse of late 2000. During this time, Valley leaders started to recast the Valley's core competency from simply being a hotbed of high-tech to one that is known for a broad, deep base of creativity and innovation. New types of skills, capacities, and community infrastructure are required for success.

New Value for Design Disciplines

One interesting implication of this shift is that people with specific training in art and design are taking their places in the high-tech workforce. More people with training in fields like product design, interactivity, user experience, Web design, animation, graphic design, digital media, game design, and brand strategy are working in high-tech as employees, contractors, or consultants.

While it should not be overstated, a range of art and design disciplines may be important to Silicon Valley in the future. A new set of art and design careers may provide an interesting alternative career path to new middle-class professional jobs. The State of California projects 45,000 new jobs in art, design, and entertainment to be created between 2002 and 2012. Two of the fast-growth subcategories are "multimedia artists and animators" and "commercial and industrial designers."

In Silicon Valley, Cogswell College promotes the "fusion of art and engineering" and helps students launch rewarding careers working on special effects, animation, scripts, music and sound in the motion picture, video gaming, and high-tech industries. Cogswell students are hired by digital entertainment companies like Electronic Arts, LucasArts, Pixar, and Industrial Light and Magic, but also by Cisco, Applied Materials, and Intel. San Jose State University is the largest provider of both art and design students in the Bay Area, with particular strengths in new media and product design. And Stanford University is developing a new "D-School"—an interdisciplinary research and education institute promoting a new kind of design thinking.

In 2003, the National Research Council documented how art and design disciplines are making substantial contributions to research and product development in the fields of computer science, networking, and communications technology.[4] This interaction between artistically creative practices and traditional technology fields is starting to surface as part of the magic mix of Silicon Valley.

New Importance of Cross-Disciplinary Teamwork

Creative breakthroughs come from an increasingly wider variety of disciplines working together. Traditionally, Silicon Valley companies have valued technical specialists. More and more, companies need specialists that respect and can work with people from other disciplines—computer scientists and engineers, for example, who can work with designers, anthropologists, and marketing experts. And, in addition to people with specialized expertise, companies also value people who transcend disciplines, people who can integrate and synthesize and strategize.

Traditionally, Silicon Valley has had a "left-brain" engineering culture—emphasizing the logical, the mathematical, the sequential, the rational, the linear. But growing competition and business shifts toward consumers, software, and services place value also on the "right brain"—the visual, the empathetic, the aesthetic, the intuitive, the simultaneous, the playful. This integration of left- and right-brain capabilities is more and more a factor for success at the individual, team, organization, and community levels.[5]

As highly creative business functions concentrate in Silicon Valley and as more business operations span the globe, more workers here find that their jobs involve managing cross-border teams, processes, and operations. At San Jose State University, the largest single provider of engineers to Silicon Valley, engineering students are being groomed not just to be technical specialists, but to manage cross-border, cross-disciplinary, cross-cultural R&D teams. There is also a new emphasis on "services engineering" as a growing profession and a new curriculum. At IBM, for example (San Jose's second largest private-sector employer), more than 60 percent of revenue now stems from client services. In contrast to the lone-star cubicle dweller of the past, successful engineers must now blend engineering expertise with people skills and business knowledge, and must work on cross-disciplinary teams—most often on site, interacting daily with the client.

New Value for the Creative Community Environment

Competing on creativity requires new attention to the community quality of life and infrastructure, the context in which creativity is nurtured and takes place. The very nature of the community—the kinds of creative outlets and the atmosphere it provides—affects the creativity of current employees, and the ability of employers to attract and develop new talent. Competition for talent will only get more intense as regions worldwide begin experiencing labor shortages in the next decade caused by the accelerating retirement of Baby Boomers.

In Silicon Valley, leaders are working to add new vibrancy and dimensions to Silicon Valley's traditional suburban amenities and ambiance in order to compete on creativity. They are doing this work even as they continue to tackle more traditional challenges such as high-cost housing and transportation infrastructure.

This means investing in vital city centers—in both downtown San Jose as the region's urban center and in many smaller community and neighborhood centers—as important locations for meetings and interaction, for entertainment and enrichment. To date, Silicon Valley has succeeded as a place despite its lackluster built environment. To compete for talent and status against other world cities, long overdue improvements in urban planning, architectural quality, public spaces, and public transportation are required and are starting to become visible.

City governments are setting new expectations for architectural quality by investing in their own landmark buildings, such as the new Richard Meier-designed City Hall in San Jose's downtown. Smaller communities like Sunnyvale, Mountain View, Redwood City, and Palo Alto have successfully revitalized a network of charming, walkable town centers with new shopping, dining, entertainment, and housing opportunities. Land use plans for key employment districts have been updated to emphasize mixed-use vibrancy and higher-density, higher-quality structures—a marked departure from the "high-tech industrial campus" vision that guided the region's recent past.

This means supporting forums and initiatives that foster conversation and collabo-ration across disciplines. Joint Venture Silicon Valley's Technology Convergence Consortium, for example, is helping to speed the convergence of three cutting-edge technological disciplines—biotechnology, nanotechnology, and information technology—by promoting new partnerships among companies, research and education institutions, and investors. On excess land at the NASA Research Park in Mountain View, universities, businesses, and local governments are working together to create the Bio*Info*Nano Research and Development Institute (BIN RDI). The Institute will provide specialized research capabilities for established and start-up companies, and will create a magnet for cross-disciplinary research talent. And the new San Jose BioCenter provides office and wet lab space, as well as a supportive business environment, to a wide variety of bioscience start-ups; for many, their core technology or business application links to information or nanotechnology.

This means valuing cultural amenities and arts education for their link to Sili-con Valley's economic future, as well as for their value to community-building. This starts with measuring what matters. Silicon Valley's *Creative Community Index* (2002, 2005) is undoubtedly the most comprehensive study of regional creativity conducted in any U.S. region. Produced by Cultural Initiatives Sili-con Valley, the *Index* affirms the very strong value that residents and employ-ees place on K–12 arts education and on their personal participation in artistic activities.

With support from the Packard Foundation, the region has made strides in restoring education in traditional visual and performing arts to the public ele-mentary schools. The challenge is to sustain this work and to layer on new skill development in design and digital media, which are important for suc-cess in the creative economy.

New, affordable live-work spaces are opening in 2006 that can accommo-date nearly 150 artists. And, while arts organizations of all sizes continue to struggle financially, examples abound of artistic excellence, strong board and staff leadership, and effective audience development initiatives, such as www.artsopolis.com.

This means raising San Jose/Silicon Valley's stature as a world cultural center and contributor. The strategy is to develop and position San Jose/Silicon Valley as North America's leading center for creativity at the intersection of art and tech-nology. This involves a new biennial international art festival, launched in August 2006, called *ZeroOne San Jose.* Other components of the strategy, which is embraced by a collaborative of civic institutions and corporate backers, are a new facility focused on art and technology exhibitions and education, and an airport public art initiative that is the largest new media curatorial program in the world.

To continue attracting and developing talent over the long haul, the region and its largest city aim to lead distinctively in art and culture, in addition to technology and entrepreneurship.

New Leadership That "Connects the Dots"

Like other communities, San Jose/Silicon Valley struggles with leadership. Locals lament the apparent lack of companies with a visible long-term commitment to advancing the region. The dynamism, mobility, and diversity of the area are all challenging, as is the sheer busyness of people struggling to integrate work and home—much less civic—life.

Yet one interesting new development is the launch in 2004 of a new regional leadership network called 1stACT Silicon Valley (re: <u>A</u>rt, <u>C</u>reativity, <u>T</u>echnology). 1stACT's vision is for Silicon Valley to be "the most creative place in the world." 1stACT is a network of influential leaders who see increasingly tighter ties between creativity, the arts, and the Silicon Valley economy. It is creating a new alignment of interests across sectors that can work together to ensure an appropriately creative local environment.

1stACT builds on some existing leadership organizations, such as American Leadership Forum Silicon Valley, Cultural Initiatives Silicon Valley, Joint Venture Silicon Valley, and the Arts Roundtable. But it marries this existing civic infrastructure to a CEO Advisory Council that includes top leaders from companies such as Cisco Systems, Adobe, Agilent, and Knight Ridder. In addition to its role in "thought leadership," this network of networks is advancing projects to diversify the arts audience base, to develop Silicon Valley's cultural identity, and to step up development of downtown San Jose as the region's creative urban center. Perhaps most important, 1stACT is setting the stage for increased corporate and community investment in arts, cultural, and creative infrastructure.

THE CREATIVE COMMUNITY OF SAN JOSE/SILICON VALLEY

Technology Companies ... that value creativity and employ people trained in artistic/design skills

Creative Service Companies ... that fuse arts, creativity, and technology to provide professional services

Creative Independents ... who earn their living using artistic and creative skills

Education Institutions ... that develop skills and qualities of a creative workforce

Cultural Institutions ... that celebrate and advance the arts, heritage, and the creative process

Support Networks ... that nurture and promote the creative economy.

Source: Collaborative Economics

Over its fifty-plus-year recent history, Silicon Valley has demonstrated remarkable resilience. With each wave of innovation and in-migration, the economy and community have adapted to weather change and sustain success. At this particular junction, civic and business leaders are working together to understand the nature of the changes taking place and how to set the stage for the next Silicon Valley. There is a clear sense, though, that the laissez-faire approach taken to economic development in the past—when Silicon Valley was the undisputed center of the technology universe—is no longer appropriate. The rise of many other city-regions around the world that are focused aggressively on technology-based economic development has shown local business and civic leaders that they can't take Silicon Valley's success for granted. The lone maverick, libertarian mindset that will always be a part of Silicon Valley's soul is being augmented by business, government, and philanthropic, education, and cultural institutions working together for mutual success. The global economy will benefit as Silicon Valley creates new sources of competitive advantage for a successful next act.

NOTES

1. The December 2005 World Knowledge Competitiveness Index ranks San Jose in California's Silicon Valley as the world's most competitive knowledge economy. The annual Index uses nineteen measures to rate 125 regions on their "knowledge competitiveness"—defined as the ability not just to create new ideas but also to exploit their economic value. These measures include R&D expenditure by business, higher education public spending, levels of employment in knowledge-intensive industries, and numbers of patents registered. The Index is produced by Robert Huggins Associates, a think-tank based in Cardiff, the Welsh capital.

2. Silicon Valley's model of wealth creation is a classic example of the "New Growth Theory" promulgated by Paul M. Romer of Stanford University. Romer argues that economic growth arises from the discovery of new "recipes" that rearrange and transform input from lower to higher value configurations.

3. See *Creative community index: Measuring progress toward a vital Silicon Valley*, produced by Cultural Initiatives Silicon Valley in 2005 (www.ci-sv.org).

4. See *Beyond productivity: Information technology, innovation, and creativity* by the National Research Council, National Academies Press, 2003.

5. For further development of this metaphor, see Daniel Pink's *The whole mind: Moving from the information age to the conceptual age*, 2005.

5

The Pipeline from University Laboratory to New Commercial Product: An Organizational Framework Regarding Technology Commercialization in Multidisciplinary Research Centers

SARA JANSEN PERRY, STEVEN C. CURRALL, and TOBY E. STUART

I n the twenty-five years since the Bayh-Dole Act of 1980, commercialization of academic research has increased significantly (Graff, Heiman, and Zilberman 2002). As a result, academic research has become more intertwined with industry. Participation by industry ranges from sponsoring specific research projects to affiliate membership in research centers such as National Science Foundation (NSF)-funded Engineering Research Centers (ERCs). Indeed, many government-funded research initiatives, such as the ERC program, are founded on the premise that multidisciplinary university and industry collaboration will enhance research productivity and foster technological advances that otherwise would not be possible.

Despite the success of the United States government's science and engineering funding initiatives, a topic that remains poorly understood is the interface between universities and industry. Often, industry leaders misunderstand how to work with universities, which, relative to for-profit companies, operate on longer timelines and can have vastly different organizational cultures, norms, and incentives. Likewise, university researchers often struggle to understand

This material is based upon work supported by the National Science Foundation under Grant No. EEC-0345195. We thank Lynn Preston and Linda Parker of the National Science Foundation for support and guidance of this research. We wish to thank Mary Sommers Pyne and Timmie Wang at the Rice Alliance for Technology and Entrepreneurship at Rice University for their assistance.

the needs of industry and the benefits corporations expect to receive from partnerships involving technology transfer (Steenhuis and Gray 2005; Thursby and Thursby 2003).

Technology transfer refers to a broad category of activities involving the translation of academic science and engineering discoveries into information that can be used by for-profit or nonprofit organizations. This may include non-commercial activities such as information dissemination through publications or research seminars. We refer to technology commercialization as the transformation of science or engineering discoveries into intellectual property, which then serves as the basis for creating new commercial products and processes (Dudley and Rood 1989). The technology commercialization concept encompasses the full process (i.e., "pipeline") of commercial activities, including invention disclosures, patent filings, licensing, and/or formation of spin-off companies.

The purpose of this chapter is to explicate the technology commercialization pipeline occurring inside multidisciplinary university research centers. The chapter is intended for two audiences. First, technology commercialization scholars can use our analysis of ERCs to supplement the field's knowledge base of the predictors of success in university technology commercialization. Second, the chapter provides insight for leaders of corporate research and development (R&D) programs concerning how to partner with universities to commercialize new technologies.

The chapter is organized into five sections. First, we provide an overview of the NSF ERC program. Next, we describe the qualitative and quantitative data collection process we used. Third, we describe general characteristics of the ERCs, including their structure, history, and organizational functioning. Fourth, we explain organizational sources of heterogeneity concerning how ERCs pursue technology commercialization. We conclude with implications for scholars of university technology transfer and implications for corporate R&D executives.

OVERVIEW OF THE NATIONAL SCIENCE FOUNDATION-FUNDED ENGINEERING RESEARCH CENTER PROGRAM

Academic research centers that involve industrial collaborations and emphasize communication across the boundaries of academic disciplines have emerged as a result of entrepreneurial awareness and a shift to applied research in universities during the past twenty years (Smilor, Dietrich, and Gibson 1993). By 1994, over 1,000 industry-university research centers had been formed in the U.S. (Roessner et al. 1998). In 1985, the National Science Foundation (NSF) launched the Engineering Research Center (ERC) program, which is the flagship scheme for federally funded support of engineering research in American universities. The program's mission is to foster national well-being and economic competitiveness by promoting university-industry collaboration to maintain and advance the nation's technological leadership.

The ERC program also focuses on educating students via interdisciplinary research opportunities and close contact with industry (Feller, Ailes, and Roessner 2002). The focus on both basic and applied research as well as their combined educational focus makes the ERC program unique among funding programs by the federal government.

An ERC provides an organizational structure that functions to systematize widely dispersed teams of collaborating researchers (Bozeman and Boardman 2003). The NSF expects an ERC to initiate broad institutional and cultural change in their host universities. Research pursued within the ERC is not simply a compilation of independent research projects; rather, it is a large, coordinated effort. Therefore, research in the ERC requires interactions among multiple disciplines and researchers. Faculty members in ERCs are encouraged to collaborate with other faculty members, disciplines, and industry researchers to accomplish their research goals. Scholars in ERCs also sometimes pursue other forms of professional activity that increase collaborations with industry, such as consulting (Mowery 1998; Smilor, Dietrich, and Gibson 1993).

Impact of ERCs

Currently there are twenty-two active ERCs. Between 1985 and 2002, the NSF funded a total of thirty-seven ERCs. The level of support from the NSF is sizeable; for example, in 2002 (the most recent data summary compiled by the NSF), the NSF allocated over $60 million to its ERCs ($2 million to $3.7 million per year per ERC). Table 5.1 presents a summary of average yearly funding inputs. The first three columns in Table 5.1 provide overall funding statistics for all ERCs and all years, such that each observation is an ERC-year (i.e., ERC #1 in 2001, ERC #1 in 2002, etc.). The remaining columns provide the averages and measures of dispersion for all ERCs during the period of our data collection, namely 2001–2005.

The outputs produced by ERCs have also been substantial. Data from 1985 to 2002 show that ERC researchers produced a total of 10,922 peer-reviewed journal articles and 9,260 peer-reviewed conference proceedings. ERCs also have produced significant intellectual property; 908 inventions were disclosed and 391 patents awarded to ERC researchers. In addition, ERCs have been the origin of ninety spin-off companies that employ 927 persons (*Engineering Research Centers Program Performance Indicators Data* 2002). Table 5.2 presents a summary of average yearly research outputs for the currently active ERCs from 2001 through 2005. As in the first table, the first three columns in Table 5.2 provide overall research output statistics for all ERCs and all years, while the remaining columns provide the averages and standard deviations for all ERCs.

As reflected in the standard deviation statistics in Tables Table 5.1 and Table 5.2, a great deal of heterogeneity, or dispersion, exists among ERCs.

TABLE 5.1. Average ERC Funding Inputs by Funding Source: 2001–2005

Source of Funding	M all ERCs, all years; N=99*	SD (all ERCs, all years; N=99)*	Min/Max (all ERCs, all years; N=99)*	2001 M(SD) (N=18)	2002 M(SD) (N=19)	2003 M(SD) (N=19)	2004 M(SD) (N=23)	2005 M(SD) (N=20)
Total Funding Received (All Sources)	$6,860,830	$3,144,802	$1,594,328/ $19,373,405	$6,561,499 ($3,353263)	$7,330,731 ($4,006,959)	$6,856,865 ($3,145,618)	$6,325,762 ($2,711,211)	$7,302,916 ($2,644,913)
NSF ERC Program	$2,899,165	$944,740	$577,331/ $5,232,401	$2,449,302 ($714,941)	$2,953,921 ($812,311)	$3,002,693 ($962,087)	$2,802,812 ($979,105)	$3,264,479 ($1,084,927)
Other NSF Sources	$133,416	$285,632	$0/ $1,162,435	$148,136 ($315,029)	$166,345 ($301,924)	$134,009 ($299,669)	$141,537 ($299,392)	$78,985 ($228,694)
Industry (US and Foreign)	$605,788	$856,538	$0/ $6,266,183	$929,236 ($1,472,483)	$724,010 ($971,422)	$501.722 ($516,562)	$433,293 ($521,887)	$499,605 ($458,467)
Federal Government (US and Foreign)	$305,942	$838,053	$0/ $6,984,500	$238,750 ($431,859)	$589,629 ($1,637,085)	$210,281 ($375,323)	$199,992 ($406,262)	$309,635 ($696,148)
State Government	$502,082	$878,486	$0/ $4,464,225	$771,325 ($1,226,710)	$566,610 ($1,122,385)	$433,617 ($643,458)	$371,866 ($625,650)	$413,253 ($699,164)
Universities (US and Foreign)	$1,068,950	$1,079,488	$0/ $6,641,388	$1,381,213 ($1,575,616)	$1,054,778 ($970,974)	$932,011 ($899,545)	$955,873 ($880,881)	$1,061,508 ($1,050,775)
Other	$42,822	$172,016	$0/ $1,300,000	$78,996 ($305,414)	$63,242 ($207,338)	$24,185 ($100,067)	$30,415 ($93,210)	$22,839 ($79,228)

Note: *In the first three columns, the mean and standard deviation were calculated using 99 observations, of which each observation was an ERC-Year (e.g., ERC #1 in 2001 and ERC #1 in 2002 are separate observations).

TABLE 5.2. ERC Technology and Knowledge Transfer Outputs: 2001–2005

Research Output	M (all ERCs, all years; N=99)	SD (all ERCs, all years; N=99)	Min/Max (all ERCs, all years; N=99)	2001 M(SD) (N=18)	2002 M (SD) (N=19)	2003 M(SD) (N=19)	2004 M(SD) (N=23)	2005 M(SD) (N=20)*
Peer-Reviewed Journal Publications	30.99	23.99	0/11	31.00 (25.23)	26.63 (17.55)	30.11 (16.73)	30.39 (27.48)	36.65 (30.21)
Invention Disclosures	6.45	7.84	0/33	6.78 (8.02)	6.32 (8.65)	6.26 (6.86)	6.83 (7.57)	6.05 (8.79)
Patent Applications	4.67	5.70	0/24	5.28 (7.19)	4.32 (7.17)	4.95 (5.15)	4.52 (4.76)	4.35 (4.53)
Patents Awarded	1.37	2.38	0/11	1.06 (2.01)	1.26 (1.76)	1.05 (1.93)	1.22 (2.63)	2.25 (3.19)
Licenses Issued	2.55	6.67	0/44	2.39 (4.80)	1.58 (4.45)	2.37 (3.29)	2.61 (8.42)	3.70 (9.75)
Spin-off Companies	0.33	0.67	0/3	.39 (.78)	.37 (.76)	.47 (.77)	.22 (.52)	.25 (.55)
Spin-off Company Employees	5.2	15.09	0/73	4.50 (12.49)	5.0 (15.30)	7.16 (16.54)	5.04 (15.86)	4.35 (16.07)
Technical Codes and Standards Impacts	0.17	0.86	0/6	—†	—†	—†	.35 (1.11)	.45 (1.47)

Note: * At time of publication, we had 2005 data from twenty ERCs of the twenty-two active ERCs, due to reporting time differences.
† indicates missing data as a result of changing reporting requirements (i.e., from 2001 to 2003, these values were not reported).

These statistics provide evidence reinforcing our view that ERCs vary widely in the way they organize themselves. These differences may be reflected in varying technological opportunity or general organizational issues; we posit they are due largely to the organizational factors discussed in the third and fourth sections of this chapter. We believe that these organizational factors explain much of the variance in research productivity among ERCs, as well as their commercialization success.

DATA COLLECTION METHODS

Our research is part of a large project funded by the NSF that is aimed at discovering best practices regarding technology commercialization occurring within ERCs. The information we present in this chapter is a result of in-depth qualitative interviews with sixty personnel from the twenty-two existing ERCs, survey data from over 800 personnel, and archival data from ERC annual reports for 2001 through 2005.

We used both quantitative and qualitative data collection methods, which can be interwoven to maximize the knowledge yield (McCall and Bobko 1990) of a research endeavor (Currall et al. 1999). "Typically, 'qualitative observation' identifies the presence or absence of something, in contrast to 'quantitative observation,' which involves measuring the degree to which some feature is present" (Kirk and Miller 1986). Qualitative methods are particularly well suited for developing a grasp of organizational phenomena where a well-established body of research literature does not exist, as is the case with organizational analyses of ERCs. Accordingly, we utilized a "two-phase" design (Creswell 2002) whereby we used qualitative methods to deepen our understanding of the organizational phenomenon under study, and to develop an accurate conceptualization of how ERCs are organized and how they operate. Subsequently, during the quantitative phase, we made use of the information generated by interviews to develop survey instruments, which were administered in November 2005. Surveys yielded quantitative perceptual and attitudinal data. We will restrict our discussion mainly to our interview and archival data; survey data will be used in our future research.

In-depth interviews with ERC personnel took place in 2005. Between January and May 2005, we visited eleven ERCs and spent at least one hour with several representatives from each ERC, including directors, industrial liaison officers (ILOs), administrative directors, education/outreach officers, and research thrust leaders. Additionally, we also spoke with faculty members and students who did not hold formal leadership positions in their ERC. We followed a semi-structured interviewing process. We used a common pool of questions and varied the order of questions depending on the answers we received and the role held by the interviewee. When granted permission from the interviewee, we recorded interviews and referred to the tape to complete our notes for each interview. Additionally, in most cases, at least two researchers participated in every interview, with one research assistant attending every interview to maintain comparability across interviews. This procedure resulted in notes from at least two perspectives, in addition to the actual audiotapes. We ensured that our interpretations from each interview were accurate and consistent across the interviewers.

We also conducted phone interviews between January and November 2005, with directors and/or ILOs in each ERC we did not visit. Therefore, using the combination of on-site and phone interviews, we spoke with at least one representative from every ERC. This thorough interview process allowed us to grasp the diversity of the ERCs and the context in which different ERCs operate. Indeed, as we will discuss later, we found that ERCs varied substantially due to different technology foci, geographic locations, and host universities.

Our conceptualization of technology commercialization outcome metrics and antecedents was informed by our qualitative data. We advocate a multifaceted approach for operationalizing commercialization effectiveness (Banner and Gagné 1995). Thus, commercialization effectiveness data were collected in terms of quantitative archival data on the number of technology transfer outputs, including invention disclosures, patent applications, patents awarded, and spin-off companies. Likewise, we took a multidimensional approach to conceptualizing the antecedents of commercialization success and their measures. The antecedents we uncovered during the interviews provided insights into the ways in which ERCs vary in their commercialization practices and success. Organizational variables were the primary determinants we found to explain variation across ERCs. We explicate these variables later in this chapter. First, however, we describe our general observations about characteristics of ERCs and how they function as organizations.

CHARACTERISTICS OF ERCs

Pipeline to Commercialization

Traditionally, academic research was viewed as a linear process, progressing from basic science to applied science to product development (Cohen, Nelson, and Walsh 2002; Croissant, Rhoades, and Slaughter 2001). Universities historically focused on the basic science component. More recently, this viewpoint has shifted as the focus to commercialization has shifted. Universities no longer restrict themselves to basic science. Further, universities now contribute to both sustaining (i.e., incremental) innovations and disruptive innovations (Rice, Leifer, and Colarelli-O'Connor 2002). Therefore, collaboration among university researchers and companies may help companies stay competitive by making incremental improvements to existing products or by leveraging technical discoveries to create completely new products. This is becoming increasingly common as universities contribute to development of new products (Cohen, Nelson, and Walsh 2002; Croissant, Rhoades, and Slaughter 2001).

Throughout their history, ERCs have emphasized both sustaining and disruptive technologies. In recent years, the ERC program management has placed an increasing emphasis on awarding grants to ERCs engaged in "pre-paradigmatic" or "transformational" research. These ERCs attempt to create new science and engineering paradigms, which can result in revolutionary technologies in new or existing industries. In line with NSF terminology, we use the term "pre-paradigmatic" to describe the technology of these ERCs. "Paradigmatic" or "next generation" research centers, on the other hand, build upon existing science and engineering models; they are focused on producing incremental technological improvements to sustain existing product lines and processes. We refer to these as "paradigmatic" technologies. Table 5.3 lists the ERCs, their demographic information, and their technology categorization as pre-paradigmatic or paradigmatic.

ERCs typically restrict themselves to activities designed to prove a technological concept or establish a limited application of the technology. These are two of the three well-defined stages of disruptive technologies (Myers et al. 2002). Proof of concept is the first stage; it shows viability of the research idea. The second stage involves establishing a limited application, which supplies the rationale for the new technology's use over the conventional approach. This is the stage at which ERCs may push the technology toward commercialization via the university's Office of Technology Transfer (OTT) or another commercialization avenue. The final stage of technological maturity, widespread commercial application, is not part of the ERC charter.

Extensive further investment must be made to bring an ERC-based technology to a stage that is ready for commercial application. Even if the university is accommodating by lowering the cost of licensure, a company licensing the technology still takes a considerable risk. According to one study, only 12 percent of development projects become commercial products, and two-thirds of those 12 percent succeed (Raine and Beukman 2002). Another study reports that 46 percent of all licensed inventions fail before product development, while 72 percent of inventions licensed at the proof of concept stage fail (Thursby and Thursby 2003).

The risk in licensing new technology is high. Typically, more funding is necessary for the technology before it can succeed as a commercial product. The ERC model does not have a built-in mechanism to advance this transition from immature technology to mature product. This is one explanation for why ERCs have varying levels of success in commercialization. In the next section, we delineate the organizational factors we believe to be central in determining the rate of transition from university research to commercial products. Scholars who study university technology commercialization can use these factors in theorizing about the determinants of commercialization effectiveness. Industry executives and leaders of agencies can use these factors to better enable collaboration and commercialization success.

TABLE 5.3.   ERC Technology Descriptions and Categorizations

ERC	Founding Year	Host University	Technology Area	Pre-Paradigmatic or Paradigmatic Technology
ERC for Environmentally Benign Semiconductor Manufacturing	1997	Univ. of Arizona	Manufacturing and Processing	Paradigmatic
ERC for Extreme Ultraviolet Science and Technology	2004	Colorado State Univ.	Microelectronic Systems and Information	Pre-Paradigmatic
Center for Neuromorphic Systems Engineering	1995	California Institute of Technology	Microelectronic Systems and Information	Pre-Paradigmatic
Mid-America Earthquake Center	1998	Univ. of Illinois	Earthquake Engineering	Paradigmatic
The Center for Enviromentally Beneficial Catalysis	2004	Univ. of Kansas	Manufacturing and Processing	Paradigmatic
ERC for Collaborative Adaptive Sensing of the Atmosphere	2004	Univ. of Massachusetts	Manufacturing and Processing	Pre-Paradigmatic
ERC for Biomimetic MicroElectronic Systems	2004	Univ. of Southern California	Bioengineering	Paradigmatic
Particle Engineering Research Center	1995	Univ. of Florida	Manufacturing and Processing	Paradigmatic
Center for Low Cost Electronic Packaging	1995	Georgia Institute of Technology	Manufacturing and Processing	Paradigmatic
Biotechnology Process Engineering Center	1995	Massachusetts Institute of Technology	Bioengineering	Pre-Paradigmatic

TABLE 5.3 (*CONTINUED*)

ERC	Founding Year	Host University	Technology Area	Pre-Paradigmatic or Paradigmatic Technology
ERC for Reconfigurable Manufacturing Systems	1996	Univ. of Michigan	Manufacturing and Processing	Paradigmatic
Integrated Media Systems Center	1997	Univ. of Southern California	Microelectronic Systems and Information	Pre-Paradigmatic
Engineered Biomaterials ERC	1997	Univ. of Washington	Bioengineering	Pre-Paradigmatic
Center for Advanced Engineering Fibers and Films	1999	Clemson Univ.	Manufacturing and Processing	Pre-Paradigmatic
ERC for Engineering of Living Tissues	1999	Georgia Institute of Technology	Bioengineering	Pre-Paradigmatic
Center for Computer-Integrated Surgical Systems and Technology	1999	Johns Hopkins Univ.	Bioengineering	Pre-Paradigmatic
Center for Power Electronics Systems	1999	Virginia Institute of Technology	Microelectronic Systems and Information	Paradigmatic
Marine Bioproducts Engineering Center	1998	Univ. of Hawaii	Bioengineering	Paradigmatic
VaNTH ERC for Bioengineering Education Technologies	2000	Vanderbilt Univ.	Bioengineering	Pre-Paradigmatic
Center for Wireless Integrated MicroSystems	2001	Univ. of Michigan	Microelectronic Systems and Information	Paradigmatic
Center for Subsurface Sensing and Imaging Systems	2001	Northeastern Univ.	Microelectronic Systems and Information	Paradigmatic

TABLE 5.3 *(CONTINUED)*

ERC	Founding Year	Host University	Technology Area	Pre-Paradigmatic or Paradigmatic Technology
Pacific Earthquake Engineering Research Center	1998	Univ. of California– Berkeley	Earthquake Engineering	Paradigmatic
Multidisciplinary Center for Earthquake Engineering Research	1997	Univ. at Buffalo	Earthquake Engineering	Paradigmatic

Strategic Planning

NSF funding of ERCs is typically awarded in approximately two five-year terms, up to a maximum of eleven years. The NSF requires that long-term strategic goals and plans be reevaluated frequently. Additionally, short-term goals and action plans help ERCs achieve their long-term goals. To facilitate this goal setting and strategic planning, the NSF introduced a tool in 1997 called the three-plane framework (see Figure 5.1).

The three-plane framework is used to align long-term and short-term planning and organizational resource allocation. The NSF requires ERCs to use the three-plane framework in all strategic planning activities. It consists of three levels, which represent the advancement of research from basic to applied. The first plane focuses on fundamental, or "basic science." The NSF expects the majority of ERC research to fall into this category. The middle plane is "enabling technologies." The research in this plane reflects the middle ground between basic research and commercial products. It brings together basic research outputs and assembles them into components that can be used to form a substantive output. The third plane is the "engineered systems" plane. This plane reflects applied research aimed at developing commercializable products or processes. Engineered systems assemble the research from the lower two planes into technologies that may be commercializable.

ERC research is oriented toward the engineered systems plane. This is fundamentally different from the traditional university research paradigm, which is more driven by curiosity than focused on translational research and engineered system deliverables. Research thrusts (i.e., groups of researchers) inside the ERC are often formed according to the three-plane framework. Sometimes a thrust maps vertically and conducts research in all three planes. Other times, thrusts are horizontal and focus on only one plane. Often,

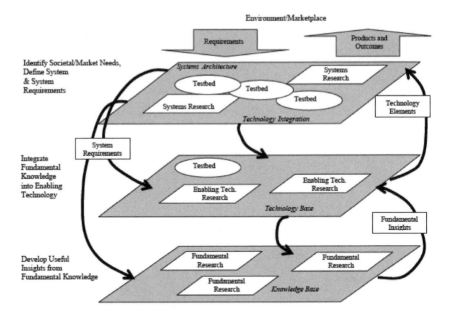

FIGURE 5.1   Three-Plane Framework

testbeds are formed as a mechanism for integrating the work of all the thrusts; testbeds most often reside in the third plane. Although ERCs vary in their implementation and interpretation of NSF mandates, the three-plane framework provides a template to which ERCs generally adhere. The framework has implications for the organizational structure of ERCs.

Organizational Structure

The strategic planning requirement is one aspect of ERCs that makes them operate as more formal organizations than academic departments, which only rarely engage in formal strategic planning. ERCs also have several leadership teams: a core leadership board, an industrial advisory board, and a scientific advisory board. The core leadership board consists of the internal ERC leadership roles: a director, an assistant director, an administrative director, an industry liaison officer (ILO), an education/outreach officer, and two to five research thrust leaders. The director and the research thrust leaders are usually faculty members, while persons with corporate and/or administrative backgrounds often hold the other positions. The industrial and scientific advisory boards are in place to provide input on the direction of research. Most member companies send one representative to serve on the industrial advisory board. This is an important role for industry partners because it allows them to provide input as well as gain access to the latest research knowledge and discoveries. Indeed, companies that maintain close ties report receiving the most benefits from their participation with the ERC (Feller, Ailes, and

Roessner 2002). Finally, the scientific advisory board is made up of key experts in the areas in which an ERC conducts research.

ERCs are hosted by one lead university and partner with several other universities. The administrative function of leadership resides at the lead university. Collaborations are often geographically dispersed around the U.S. The ERC educational programs also focus on diversity and outreach, which encourages collaboration with historically minority-oriented universities. ERCs also give early exposure to elementary- and high school-aged students, in an effort to recruit the next generation of scientists and engineers.

In addition to providing general requirements for the basic organizational structure, the NSF closely oversees the organizational functioning of ERCs. Each year, an NSF review panel visits each ERC for several days to provide feedback on successes and areas requiring improvement. ERCs hold one or more retreats per year to plan for the site visit, engage in strategic planning, and revisit their overall goals. Advisory boards and faculty attend these retreats. The leadership teams and students from every ERC also attend the annual ERC conference in Washington, D.C., hosted by NSF, at which they consult with other ERCs to share best ERC practices and receive from NSF the strategic focus of the overall ERC program for the upcoming year. In addition to the formal gatherings, each ERC holds informal events such as consulting days where lawyers or entrepreneurs provide information on commercialization activities, students present their research to industry representatives, lunches are held to share research progress, and other opportunities are provided for the parties to share information.

In sum, ERCs are organizational entities themselves. They are usually dispersed over large geographic areas and function as virtual research organizations. Because of their unique features, ERCs make ideal candidates for study by organizational researchers, especially in the area of technology commercialization and industry-university collaboration.

Industry Relationships

A hallmark of the ERC program is significant involvement by industry. This is reflected in the following statement regarding the ERC Program management's vision for ERCs: "Thus, ERCs provide the intellectual foundation for industry to collaborate with faculty and students on resolving generic, long-range challenges, producing the knowledge needed to ensure steady advances in technology and speed their transition to the marketplace, while training graduates who are more effective in industry" (*Engineering Education and Centers Division, document 00-137a* 2003). ERCs are similar to other academic research centers funded by the NSF, yet they are unique because of their focus on development of engineered systems rather than solely on basic research. They are also unique in their focus on collaboration with multiple disciplines, universities, and industrial partners.

Feller and colleagues (Feller, Ailes, and Roessner 2002; Feller and Roessner 1995; Roessner et al. 1998) have extensively studied the industry partner perspective in ERCs. They have found that industry receives many benefits from participation in ERCs, especially when they are closely engaged and maintain frequent contact. Perhaps surprisingly, partner firms do not place the greatest value on the acquisition of intellectual property but, rather, place most value on the knowledge transfer outputs from an ERC in the form of students graduating and access to cutting-edge science. One interviewee attributed this to the fact that, if the university maintains intellectual property rights, then no other competitor can own it, and the university can fight the battle of ownership with others instead of the industry partner fighting that same battle with its competitors.

From the industry perspective, the benefits of an affiliation with an ERC are made possible because substantial NSF funding helps establish a successful research infrastructure without requiring large investments from industry. The research infrastructure brings key faculty together and provides a forum for collaboration on a common integrative theme that spans departments, disciplines, and institutions. Because of these benefits, companies often pay membership dues and participate without requiring detailed economic justification for their participation. Most feel they receive many benefits, but those benefits are not easy to translate into bottom-line figures; often they are intangible and hard to quantify (Feller, Ailes, and Roessner 2002).

Herein lies a significant challenge for ERC sustainability after the NSF funding ends. When government funding ends, ERCs often look to industry for financial sustainability. However, because industry cannot easily quantify the payoffs of their participation, they may not invest more than nominal fees in an ERC. Sponsored research is easier to justify because it is tied to a specific deliverable. But sponsored research income alone cannot fund an ERC and the overhead associated with it. Indeed, membership dues for an ERC are unlikely to financially sustain an ERC. Therefore, Feller and colleagues argue that the ERC is a financially fragile organization because, without continued government support, it cannot maintain its industry partnering model and success (Feller, Ailes, and Roessner 2002).

In our examination of ERCs, we uncovered this pattern, which may reflect the lack of sustainability know-how on the part of government leaders. Some ERCs have begun demonstrating success at self-sufficiency, but the current ERC model itself is not conducive to successful sustainability through industry funding. This is one of the biggest topics of concern among ERC personnel today. Our research indicates that the most successful ERCs tend to view themselves as start-up companies or enterprises. However, often ERC leaders have academic roots, not entrepreneurial roots, which make them unfamiliar with managing the ERC as an enterprise. The NSF paradigm for sustainability advocates income from government, university, and industry sources. The government's emphasis on industry funding, however, is often overly optimistic. Another perspective worthy of consideration is the inclusion of endowments (private donors) as a category of financial support for ERCs. Adding private donors to the list of possible funding sources is a promising possibility. Overall, it seems that the ERC organization may indeed be fragile when the time comes for financial self-sustainability.

ORGANIZATIONAL FACTORS IMPACTING COMMERCIALIZATION OF UNIVERSITY TECHNOLOGIES

In this section, we use preliminary findings from our study of ERCs to shed light on sources of heterogeneity across ERCs with respect to technology commercialization.

Organizational Porosity

Porosity describes the flow of resources (information, people) across organizational boundaries. A research organization can be porous in two ways: intraorganizational porosity and extraorganizational porosity. Intra-ERC porosity is the collaboration that occurs within the ERC, among individual investigators and research teams, and across academic disciplines. Many people we interviewed emphasized the importance of *interdependent* research teams in collaboration across these boundaries. Also mentioned was the important role of testbeds in integrating research teams toward a common goal, which therefore facilitates collaboration. A higher level of collaboration appears to encourage more commercialization productivity because of the ability to tackle complicated applied problems that require input from various research thrusts and different academic disciplines.

Extra-ERC porosity involves the collaboration and communication between the ERC and its affiliates, such as its industry partners and other universities. Again, the importance of frequent communication with industry partners was emphasized throughout our interviews. More interaction between the ERC and the industry appears to lead to more commercialization productivity because of input about market demand from industry experts or entrepreneurs (Feller, Ailes, and Roessner 2002).

An important antecedent of intra- and extra-ERC porosity is the level of trust among the parties. Indeed, trust has been suggested as an important indicator of the degree to which universities commercialize technology (Gopalakrishnan and Santoro 2004). Intra-ERC trust involves trust among researchers, within and across research teams and across the boundaries of traditional academic disciplines. Extra-ERC trust reflects the level of trust ERC personnel have with industry partners, and vice versa. In the ERC context, increased trust likely leads to more porosity because of enhanced information flow, more cooperative contractual agreements, and decreased transaction costs (Inkpen and Currall 2004), all of which are expected to lead to more commercialization activity.

We view porosity as a central consideration when analyzing commercialization effectiveness. By the nature of collaboration, a high level of communication is required among researchers and between industry and universities. Therefore, trust formation must be an important requirement in enhancing collaboration. From our interview results, we believe that ERCs that have high levels of communication and collaboration across boundaries are most successful in terms of shorter commercialization timelines and success in commercialization activities.

ERCs' Managerial Structure

Managerial structure is defined by three facets: hierarchical levels, centralization, and formalization. Some authors posit that because technological innovations must start with those people with technical competency, technology commercialization occurs most successfully when a research organization is organic (i.e., flatter structure, decentralized, and informal). Yet, the question of structure is complex because the mere existence of a multi-disciplinary research center such as an ERC demands a certain degree of centralization, formalization, and hierarchy (Bozeman and Boardman 2003).

Bozeman and Boardman (2003) make several recommendations for organizing research centers, such as aligning reporting lines to the unique culture of the university, delineating responsibilities between an administrative director and research director, fostering creative competition by offering seed money for promising grant proposals, and cultivating center-wide relationships through gatherings and communications that include faculty and industrial partners. These efforts can be challenging to implement when a research center attempts to partner with multiple universities, especially when they involve the work of researchers who have little visibility with their respective university leaders. Leadership buy-in from all universities involved is very important; this may be harder to achieve if organizational structures and processes are mismatched among the universities and the ERC.

ERCs unavoidably introduce some degree of bureaucracy. In our interviews, many faculty members complained about the administrative burdens of ERC involvement. However, most acknowledged that the benefits of involvement in an ERC outweigh the burdens. Nevertheless, it is important to recognize the added pressures ERC researchers must endure to pursue ERC-related research. To foster first-rate academic research, as well as technology commercialization activities, ERC leaders must minimize the administrative load experienced by ERC faculty members.

Reputation of an ERC's Host University

ERCs based in universities with favorable reputations are more likely to commercialize successfully and more frequently. This may be because industrial partners gravitate toward relationships with universities perceived to produce relevant research for markets (Bradshaw, Munroe, and Westwind 2005; Di Gregorio and Shane 2003; Sine, Shane, and Di Gregorio 2003; Stuart 2000). Further, universities with strong reputations attract faculty with successful track records in cutting-edge research and industry collaboration. Often these faculty members secure industry contacts that greatly benefit the ERC. Therefore, both the university's reputation and the reputation of its faculty members mutually reinforce an ERC's ability to succeed with commercialization efforts.

Commercialization Infrastructure of an ERC's Host University

Support systems are an important aspect of commercialization activity. Reasonable patent protection, ownership, and licensing policies are crucial to success in commercialization (Gopalakrishnan and Santoro 2004). Office of Technology Transfer (OTT) policies should encourage commercialization by reducing risk to the licensing companies and reducing risk to the inventors. The ILO role is a featured role in ERCs that gives further infrastructure support to faculty interested in commercializing. The ILO acts as an important boundary person who can negotiate relationships between universities and industry, as well as personally market technologies to company contacts. The ILO also may act as a liaison to the OTT to ease the burden of the commercialization process on faculty. We found that ERCs with active, marketing-oriented ILOs tended to be more successful than ERCs whose ILOs focused primarily on industrial membership recruitment and partnering. Moreover, if the ILO is successful, his or her work can further drive the entrepreneurial culture that supports commercialization (Kassicieh, Radosevich, and Umbarger 1996).

In addition to the industrial liaisons provided by the ERCs, our interviews uncovered other features of ERCs that encourage translation of research into commercialized technologies such as dedicated support staffs, state-of-the-art research facilities, and access to dedicated office equipment. One interviewee called this a "universe of support" and suggested that it is vital to achieving commercialization success.

Industry Targets and Technology Area

The commercialization process differs across industries. Some industries include companies that are more receptive to university technologies (Hanson 1995). Prototypes are required before companies consider licensing; yet in certain industries, such as the life sciences, it takes longer to develop a product for testing (Hsu and Bernstein 1997). Further, life sciences are an example of an industry that must surpass regulatory hurdles before commercialization can occur. Nevertheless, the life sciences are a source of sizeable commercialization activity, accounting for up to two-thirds of total patent activity (Fisher and Atkinson-Grosjean, 2002; Mowery 1998). Companies in the semiconductor and electronics industries also are very active in commercialization (Mowery 1998). The upshot is that if an ERC is involved in an industry with high commercialization rates, the ERC is expected to have more commercialization success.

In addition to the type of industry and technological focus, other industry factors were discovered in our interviews. Many respondents mentioned the need to manage expectations with industry, and most mentioned the clash of timelines between industry and university research. Other interviewees mentioned the need to maintain close contact with industry in order to receive timely feedback and input; this maintains the commercial relevance of university research. But most ERC personnel also mentioned that restricting industry partners to an advisory role, and not a management role, is a key condition to making collaboration successful.

Many respondents emphasized the concern that ERCs can truly only be the "R" in R&D. In other words, ERCs are only equipped (i.e., in regard to funding, facilities, researchers, timelines, etc.) to do the initial research on technology, but not actual product development. The testbeds of the ERCs seem to be the best place for optimal collaboration, because they conduct applied aspects of the research. Therefore, many interviewees said that a match between the testbeds and the interests of industrial companies leads to successful collaboration. Finally, many interviewees asserted that corporations need to learn how to work with universities and that ERCs need to learn how to work with corporations. This goes back to a fundamental premise of this chapter: an understanding of the interface between industry and university must be achieved to increase commercialization and collaboration success.

In summary, we have observed several antecedents to commercialization effectiveness. We believe that porosity of organizational culture, managerial structure, university reputation, university infrastructure, and industry/technology area are important when explaining commercialization heterogeneity across ERCs. In the next section, we apply these observations to universities more broadly to contribute to scholarly discourse regarding the factors that make universities successful in commercialization. An increased understanding of these issues also will facilitate improved corporate-university collaborations.

IMPLICATIONS FOR TECHNOLOGY TRANSFER SCHOLARS AND CORPORATIONS

ERCs provide an intriguing context in which to study university technology commercialization. They are enterprising and industry-focused; therefore, they reflect trends in many universities toward greater entrepreneurial activity. We have inductively explored the organizational functioning of ERCs, as a specific example of how universities engage in technology commercialization. Our initial findings indicate that heterogeneity exists across ERCs with regard to technology commercialization practices and organizational functioning.

Scholars interested in university-industry technology commercialization can use the variables we uncovered in exploring technology commercialization heterogeneity across ERCs to develop new conceptual models of university technology commercialization. Further, ERCs hold many interesting opportunities for learning about universities, especially entrepreneurial universities, which are committed to technology transfer. The technology commercialization focus of ERCs will be interesting to scholars who study technology transfer because the data available is vast and the record keeping is unsurpassed in other types of academic research organization.

As always, causation is difficult to determine without a complex, controlled, longitudinal study design. Even then, it is often difficult to disentangle cause and effect of organizational phenomena. For instance, do our organizational variables invariably cause technology commercialization, or could the relationship be bi-directional? Although we have made extensive use of longitudinal qualitative and quantitative data, we cannot conclude with certainty the direction of relationships discussed here. These concerns are valid, and we acknowledge this limitation of the present study, especially in these preliminary findings.

Individuals in companies that collaborate with universities should assess where a university falls on the spectrum of each of the organizational variables we discussed above. How encouraging is the university toward faculty members commercializing their research outputs? How bureaucratic, hierarchical, and centralized is the university and/or academic department hosting the research efforts? Is it organic and flexible, allowing the researchers to collaborate freely? What resources are provided in the way of commercialization support, and what policies are in place in the university's OTT? How open are the communication lines between the university and its industry partners? These are important questions to ask about a university with which a company aims to collaborate.

Answers to these questions will assist company executives in understanding barriers and concerns of the university. Partnerships must be positioned in terms of long-term research benefits for the university (e.g., sponsored research funding and access to industry experts), which may motivate the university to relax the terms of licensing demands. If an industrial liaison exists, they can serve as boundary communicators because they understand both university and industry concerns. Therefore, it may be advisable for companies to initiate contact with ILOs rather than university technology transfer professionals or individual faculty members.

Scholars and industry leaders alike can learn from the findings presented in this chapter in regard to how universities participate in technology commercialization. The preliminary findings on ERCs that we presented in this chapter are intended to help scholars better explore commercialization success in universities. We hope that lessons learned from our study of ERCs are useful to industrial R&D executives in building effective commercial relationships with multidisciplinary academic research centers.

REFERENCES

Banner, D. K. and Gagné, T. E. (1995) *Designing effective organizations: Traditional & transformational views.* Thousand Oaks, Calif.: Sage Publications.

Bozeman, B. and Boardman, P. C. (2003) Managing the new multipurpose, multidiscipline university research centers: Institutional innovation in the academic community. Arlington, Va.:Transforming Organization series from IBM Center for the Business of Government.

Bradshaw, T. K., Munroe, T., and Westwind, M. (2005). Economic development via university-based technology transfer: Strategies for non-elite universities. *International Journal of Technology Transfer and Commercialisation* 4 (3): 279–301.

Cohen, W. M., Nelson, R. R., and Walsh, J. P. (2002) Links and impacts: The influence of public research on industrial R&D. *Management Science* 48 (1): 1–23.

Creswell, J. W. (2002) *Research design: Qualitative, quantitative, and mixed methods approaches* (2nd ed.). Thousand Oaks, Calif.: Sage Publications.

Croissant, J. L., Rhoades, G., and Slaughter, S. (2001) Universities in the information age: Changing work, organization, and values in academic science and engineering. *Bulletin of Science, Technology, & Society* 21 (2): 108–118.

Currall, S. C., Hammer, T. H., Baggett, L. S., and Doniger, G. M. (1999) Combining qualitative and quantitative methodologies to study group processes: An illustrative study of a corporate board of directors. *Organizational Research Methods* 2: 5–36.

Di Gregorio, D. and Shane, S. (2003) Why do some universities generate more start-ups than others? *Research Policy* 32: 209–227.

Dudley, L. S. and Rood, S. A. (1989). Technology commercialization: Combining public and private. *Policy Studies Journal* 18 (1): 188–202.

Engineering Education and Centers Division, document 00-137a. (2003) Arlington, Va.: National Science Foundation.

Engineering Research Centers Program Performance Indicators Data. (2002) Arlington, Va.: National Science Foundation.

Feller, I., Ailes, C. P., and Roessner, D. (2002) Impacts of research universities on technological innovation in industry: Evidence from engineering research centers. *Research Policy* 31: 457–474.

Feller, I. and Roessner, D. (1995) What does industry expect from university partnerships? *Issues in Science and Technology* 11 (1): 80–84.

Fisher, D. and Atkinson-Grosjean, J. (2002) Brokers on the boundary: Academy-industry liaison in Canadian universities. *Higher Education* 44: 449–467.

Gopalakrishnan, S. and Santoro, M. D. (2004) Distinguishing between knowledge transfer and technology transfer activities: The role of key organizational factors. *IEEE Transactions on Engineering Management* 51 (1): 57–69.

Graff, G., Heiman, A., and Zilberman, D. (2002). University research and offices of technology transfer. *California Management Review* 45 (1): 88–115.

Hanson, D. (1995) Study confirms importance of federal role in technology commercialization. *Chemical & Engineering News* 73 (48): 16.

Hsu, D. H. and Bernstein, T. (1997) Managing the university technology licensing process: Findings from case studies. *Journal of the Association of University Technology Managers* 11: 1–18.

Inkpen, A. C. and Currall, S. C. (2004) The co-evolution of trust, control, and learning in joint ventures. *Organization Science* 15: 586–599.

Kassicieh, S. K., Radosevich, R., and Umbarger, J. (1996) A comparative study of entrepreneurship incidence among inventors in national laboratories. *Entrepreneurship Theory and Practice* 20 (3): 33–49.

Kirk, J. and Miller, M. L. (1986) *Reliability and validity in qualitative research.* Beverly Hills, Calif.: Sage Publications.

McCall, M. W. and Bobko, P. (1990) Research methods in the service of discovery. In Dunnette, M. D. and Hough, L. M., eds. *Handbook of industrial and organizational psychology* (2nd ed.). Palo Alto, Calif.: Consulting Psychologists Press. 381–418.

Mowery, D. C. (1998) Collaborative R&D: How effective is it? *Issues in Science and Technology* 15 (1): 38–44.

Myers, D. R., Sumpter, C. W., Walsh, S. T., and Kirchhoff, B. A. (2002). A practitioner's view: Evolutionary stages of disruptive technologies. *IEEE Transactions on Engineering Management* 49 (4): 322–329.

Raine, J. K. and Beukman, C. P. (2002) University technology commercialisation offices—a New Zealand perspective. *International Journal of Technology Management* 24 (5/6): 627–647.

Rice, M. P., Leifer, R., and Colarelli-O'Connor, G. (2002) Commercializing discontinuous innovations: Bridging the gap from discontinuous innovation project to operations. *IEEE Transactions on Engineering Management* 49 (4): 330–340.

Roessner, D., Ailes, C. P., Feller, I., and Parker, L. (1998) How industry benefits from NSF's engineering research centers. *Research Technology Management* 41 (5): 40–44.

Sine, W. D., Shane, S., and Di Gregorio, D. (2003) The halo effect and technology licensing: The influence of institutional prestige on the licensing of university inventions. *Management Science* 49 (4): 478–496.

Smilor, R. W., Dietrich, G. B., and Gibson, D. V. (1993) The entrepreneurial university: The role of higher education in the United States in technology commercialization and economic development. *International Social Science Journal* 45 (1): 1–11.

Steenhuis, H.-J. and Gray, D. O. (2005). Strategic decision-making in publicly funded innovative organisations: An exploratory study. *International Journal of Technology Transfer and Commercialisation* 4 (2): 127–147.

Stuart, T. E. (2000) Interorganizational alliances and the performance of firms: A study of growth and innovation rates in a high-technology industry. *Strategic Management Journal* 21: 791–811.

Thursby, J. G. and Thursby, M. C. (2003) University licensing and the Bayh-Dole Act. *Science* 301 (5636): 1052.

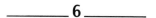

6

Bringing University Technology to the Private Sector

LUIS R. MEJIA and KIRSTEN LEUTE

B y any industrialized-country comparison, the volume of research conducted by U.S. universities is immense, surpassing $29 billion annually.[1] That research, funded mostly by U.S. taxpayers, has helped the U.S. retain its lead as the world's top innovator[2] and, in comparison to other government expenditures, has arguably provided a high yield back to the treasury and taxpayer. According to the Association of University Technology Manager's Annual Licensing Survey, over 2,500 new products based on university-licensed technology were launched in the five-year period from 1998 to 2003.[3] Thus, while universities create new knowledge and technologies in the course of their research missions, it is ultimately up to U.S. industry to find ways to use those research results to make commercially viable products.

In today's globally competitive environment, it is more imperative than ever that U.S. industry increase the rate of utilization of university-based research results. In order to do that, U.S. industry has to banish the N-I-H ("not-invented-here") syndrome from its corporate mindset. Freeing themselves from the N-I-H syndrome, companies can leverage university technology to bring new and better products to the market faster and cheaper than if they continue to try to innovate all their product pipeline needs themselves.

Indeed, companies do not have to look hard to find clear evidence that commercializing university-spawned technologies can be an astute corporate strategy. Some of the most important commercial products of the twentieth century owe their success to the university researchers who gave them

TABLE 6.1. Evolution of University Technology Transfer Parameters

	1998	2003
University Research Budget	$24.4 Billion	$38.5 Billion
University Patent Applications Filed	4,808	7,921
University License Income	$0.725 Billion	$1.3 Billion

technological birth. In the field of electronics, the driving force behind the information age, companies such as Hewlett-Packard, SUN Microsystems, and MIPS Computers all demonstrated that university-based technologies could be used to leverage successful commercial products into successful corporate enterprises. In the field of medicine, the invention of gene splicing has enabled the production of blockbuster drugs such as Remicade (for rheumatoid arthritis), Epogen (for anemia), and TPA (for stroke and heart attacks). In the information technology sector, Yahoo! and Google have parlayed their university-hatched creations into Internet juggernauts.

For corporations seeking to exploit university ideas, all trends are favorable. First, the government has recognized the importance of university research to the national economy and has continued to increase research budgets in many areas. Second, universities have recognized the value of intellectual property, not only to further their research and education missions, but also to help transfer technologies to industry for society's use and benefit. Table 6.1 highlights these positive trends.[4]

Assuming that companies have unmet technological needs, which is probably an understatement, ready sources of new innovations are very likely sitting in the labs of some of the world's finest research laboratories—that is, the labs of the premier U.S. universities.

The purpose of this chapter is to present an overview of how the technology transfer process works at U.S. universities (using Stanford University as an example, where appropriate), with the ultimate goal being to help companies understand the relevant processes so that they can be prepared to exploit the research and innovations that are made available for public use and benefit.

INNOVATION THROUGH UNIVERSITY PARTNERING

One of the organs that make up the Silicon Valley creature, Stanford has long had associations with industry players. From its industrial park to conferences to student start-ups, Stanford often partners with industry to achieve its mission of education and research. Although goals of universities and companies differ, both can benefit from working together without compromising their missions.

The Industrial Affiliates Programs at Stanford support "transfer of knowledge into society and dialogue between academia and industry."[5] A variety of affiliates programs exist in different academic disciplines, including the Center for Integrated Systems, Design, and Stanford Medical Informatics. Industry subscribers pay an annual fee to be members of the program. The fee is used to further research in the particular academic area. The subscribing companies are invited to the program's conferences detailing the ongoing cutting-edge research, and in some cases companies are allowed to place visiting researchers in a Stanford academic lab in order to promote the exchange of ideas.

Other means of academic-industrial exchange include local, national, and global conferences, companies hiring graduates, journal publications, collaborations between researchers, sponsored research, faculty consulting, and licensing of inventions created at the university. A university's mission of education and research is of utmost importance in any of these interactions; therefore, any partnership is governed by openness and freedom in research. University researchers are guaranteed the right to publish and share their results.

Despite some recent criticism, universities cannot cut themselves off from industry. Isolation keeps minds closed to possibilities, and in the ever-expanding global world we live in, working in a vacuum will stifle innovation. A university may not have the same objectives as a company, but industry and nonprofit institutions can still find ways to work together that keep their missions and goals intact and without conflict while benefiting universities, companies, and the public.

The Bayh-Dole Act of 1980 (Public Law 96-517, The Patent and Trademark Law Amendments Act, along with Public Law 98-620, Trademark Clarification Act of 1984) granted universities the right to take title to inventions created using funds received from the Federal government. Previously, the Federal government had held title to such inventions, unless the university successfully petitioned for the title. However, few of the government-owned inventions were licensed and developed into products.[6] Senators Birch Bayh (D-Indiana) and Robert Dole (R-Kansas) sponsored the legislation in part to help stimulate the U.S. economy. Under the Bayh-Dole Act, universities are to:

- grant the government a royalty-free, nonexclusive license;
- show a preference for licensing inventions to small business;
- share any income with inventors;[7]
- require a company taking an exclusive license to substantially manufacture the licensed technology's product in the United States;[8] and
- use the income to support further academic research, as well as comply with other conditions.[9]

University licensing offices were founded to combine the mission of the university, the purpose of the Bayh-Dole Act, and the promotion of public welfare. For example, the stated mission of Stanford University's Office of Technology Licensing (OTL) is:

To promote the transfer of Stanford technology for society's use and benefit while generating unrestricted income to support research and education.

Accordingly, the OTL works with industry to bring lifesaving and life-enhancing technologies to the public while ensuring the values and goals of the university. Most technology transfer offices in the United States have similar objectives, but may also include local economic development as part of their mission.

THE UNIVERSITY TECHNOLOGY TRANSFER PROCESS—WORKINGS AND PHILOSOPHIES

Inventions Disclosures

This is where is all begins—a simple form that provides the essential facts of a new invention (see http://otl.stanford.edu/inventors/disclosures.html for an example). Most, if not all, companies have invention disclosure forms to document whenever an employee has a new idea. One primary difference between university and industry invention disclosures (aside from the fact that industrial disclosures are many more pages long) is that the submission of a new invention is generally a voluntary act at universities. In other words, university researcher are not necessarily trying (nor are they obligated) to invent anything when they embark on a particular research project. They are, for the most part, driven by intellectual curiosity to do basic research—to explore scientific phenomena, model the physics of a new material, study novel computer architectures, or understand the inner working of a living cell, for example—and then to publish the results of that research.

Because a researcher's job isn't necessarily to invent, the invention disclosure process has to be simple; it cannot be viewed as a hindrance to the research. The invention disclosure form asks for very basic information:

1. The title of the invention
2. The names of the inventors
3. When the invention was first conceived/reduced to practice
4. The dates of any external disclosures
5. The sponsors of the research that led to the invention
6. A description of the invention and the problem it solves

A critical aspect of managing university-based intellectual property is that professional managers are needed to shepherd inventions from the lab to the marketplace. It is these people who take the brief information provided to them at the time the invention disclosure is submitted and decide an appropriate course of action. The first action taken by the Intellectual Property (IP) managers is to meet with the inventors to obtain more information about the invention. Meeting with inventors is a more efficient way to get relevant and important information about an invention (shortcomings, for example) because inventors are more likely to divulge such information to a person rather than writing it down.

An important note to make here is that the process of patenting and licensing an invention can be a very personal experience to an inventor. After all, the invention is his baby, and he wants it handled properly. So, the first inventor meeting serves not only to get vital details on an invention, it serves as a way for the IP manager to build rapport with the inventor. As we will see, maintaining rapport with inventors will be helpful to the success in the invention's commercial exploitation.

The first inventor meeting also provides a good opportunity for the IP manager to explain the licensing process to the inventor. It is important that inventors understand the process so that their expectations can be appropriately set. Because only a small fraction of inventions are ever licensed (and an even smaller fraction actually become commercial products), inventors should understand the difficulty of finding a licensee and then having a product make it to the launch stage.

Evaluation

Thankfully, university researchers are generally not bashful people. They usually are quite prolific when it comes to submitting invention disclosures on ideas they think might be patentable. Therein lies a difficulty. Universities are not initially concerned about whether inventions are patentable. Rather, they are mostly concerned with whether inventions are licensable.

In other words, just because an invention is patentable doesn't mean that it is commercializable. Furthermore, even if an invention is commercializable, that doesn't mean there is a big enough market to interest a company in making a substantial investment in developing the product. So, in other words, as with venture capitalists, university IP managers do not invest in technologies, they invest in markets. So, it is again important to emphasize these distinctions to the inventors during the evaluation process.

University inventions, almost by definition, are going to be early-stage ideas. The nature of basic research implies that researchers are looking for fundamental results, not merely minor improvements or advances (although that does happen too). A fundamentally new invention is fantastic from a patentability point of view, but more often than not the invention is too far from commercial viability to be of immediate interest to most companies. Sometimes, it's not even clear what markets the invention might be useful for.

So, how does a university evaluate a new idea? One typical place to start is to fill out an evaluation checklist (see the Appendix for an example). Completing the form is one way to make a quick assessment of important attributes of the invention. It is also helpful to use aspects of this list to engage the inventor in the evaluation process—it shows that the focus of the evaluation is on the invention's *licensability*, not the *science*, per se.

If there is an urgent need to beat a publication bar, and if an invention looks promising on the checklist, it might be a reasonable risk to proceed to file a patent application. However, more often than not, it is beneficial to gather more data—that is, feedback from industry—before making a filing decision.

At this point, it is worthwhile to reemphasize that the objective for universities is to out-license its inventions. Companies, on the other hand, usually patent their inventions for defensive purposes (as protection from competitors), not for generating income from out-licensing. Another way to look at this is to recognize that the university, unlike a company, does not make products to sell. Because universities are not competing with companies in the sales of products, the university can more easily solicit feedback from companies about whether the company might have a commercial interest in an invention.

As mentioned earlier, one of the distinguishing characteristics of universities is that they, by policy, make the results of their research public, usually through a peer-reviewed publication, dissertation, or academic conference. Conversely, companies usually keep the results of their research cloaked in secrecy. University publications have pros and cons from a technology transfer perspective. With regard to the evaluation of an invention, publications serve as an effective way to get feedback from industry. Because of the public disclosure, there is usually no need for confidentiality agreements, so companies can freely evaluate an invention that is described in the publication.

The feedback obtained from companies, along with the assessed licensability (using the checklist), is then used to determine what to do with the invention. This process has several possible outcomes:

1. Let the invention incubate further in the lab (assuming there hasn't been a publication);
2. File a patent application; or
3. Let the invention fall into the pubic domain.[10]

The evaluation of early-stage inventions is often a difficult thing to do. Inventions are sometimes ten to twenty years "ahead of their time." In those cases, the IP manager must use his or her best judgment as to whether the potential license income of an invention is high enough to justify the expense of patenting an invention and then sitting on the patent for a long time without any income coming in to pay for the patent expenses and overhead. Figure 6.1 illustrates the relationship between income and overhead, specifically, that it took Stanford OTL about fifteen years to get to breakeven.

Marketing

In order to encourage development of university inventions, Bayh-Dole grants universities the right to obtain patents on inventions made at least in part by federally funded research. OTL's mission states that it will "promote technology for society's use and benefit," and the Bayh-Dole Act stipulates the promotion of "the utilization of inventions arising from federally supported

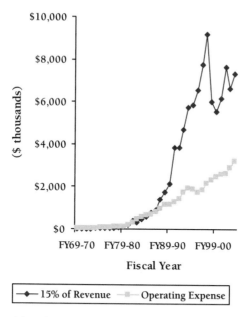

FIGURE 6.1. Financial Performance of Technology Transfer Offices

research or development."[11] As part of the evaluation process, OTL contemplates the best way to move the invention forward toward the production of a beneficial product.

Our main licensing strategy decision, then, is to decide whether the invention is best developed through nonexclusive licensing or exclusive or field exclusive licensing. An invention that can already be used by companies, would be used in research, or requires little time, effort, and money in order to turn out a product would more likely be licensed nonexclusively to multiple companies. Examples of technologies that would typically be licensed nonexclusively are research reagents, research tools, software, and processes.

On the other hand, inventions that require much time and development before a product is made, such as a compound for cancer treatment, would likely require an exclusive license as a company would not invest the time and money in research and product development without the chance to have some market exclusivity. In certain cases, even if a company is requesting an exclusive license, if the technology does not need much development and should be broadly licensed to ensure access to all players, Stanford will choose not to grant an exclusive license. The flip situation can occur when a company takes a nonexclusive license to a technology simply to have access, but then does not develop the technology and bars any other company from taking an exclusive license. Stanford includes diligence clauses in both its exclusive and its nonexclusive licenses to help ensure development of the technologies. With exclusive or nonexclusive licensing, the end goal is the same—the technology is developed and is used to assist society.

Before Stanford licenses an invention exclusively, OTL will attempt to find the best company to license, develop, market, and sell the invention. Finding that company, or companies in the case of field exclusive licenses, is one of a technology transfer office's main challenges. Stanford's OTL receives over 350 invention disclosures per year. Multiply this by all the major research universities and institutions in the world and compare with the number of existing companies and start-ups created each year, and it is obvious that companies could not undertake all of the technologies available.

The main route through which university technology transfer offices find companies to license and develop their technologies is the technology's inventors. These researchers often know many of the players in the field, have connections within companies through colleagues or sponsored research or collaborations, meet other researchers at meetings, or are approached by companies due to publications or seminars. Despite the fact that researchers are the main conduit for technology licenses, Stanford's OTL still "markets" every invention it seeks to license.

Why does OTL do this? First, part of our mission and the Bayh-Dole Act's guidelines are to find the best company possible to develop the technology. If we only look at companies that our researchers know, we may have a very limited view. Second, choosing a company simply because our researcher has a connection with that company is rife with conflicts. As an academic institution, we cannot funnel research results into a company.

In order to find other possible licensing candidates, Stanford's OTL takes a few steps. First, it creates an abstract regarding the technology that includes a written description of the technology, its advantages, the applications, the researchers involved, and notations of any publications, including patents. This abstract is posted on the OTL Web site, which is searchable.[12] OTL then conducts a search for companies that might be interested in the technology. OTL's database is populated with numerous contacts accumulated over many years, and each of these marketing contacts has keywords associated with him or her. A list of contacts can be pulled up simply by choosing a certain keyword associated with the technology. OTL will also search for new contacts through databases and web sites. The abstract is then sent to the database and search contacts by e-mail, fax or post.

A journal article or speech by the technology's inventor is often the best form of marketing a technology can receive. Recognition of a technology's importance by peer review and publication leads to a lifted view of the technology in industry's eyes as well. In some cases, OTL will wait for the release of a publication before marketing an invention.

No matter what the invention, OTL will usually market it before a patent is issued. It may market a single invention multiple times over the lifetime of the invention as new data is discovered, industry players change, new applications of the invention are found, and patents are published or issued. When OTL first markets an invention that is invented under an industrial affiliates program, mentioned earlier in this chapter, it will contact all members of the program about the technology.

As the biotechnology sector in general licenses more technologies from universities than any other industrial sector, biotechnology companies often have a contact or group that deals only with universities and their offerings. Biotechnology companies respond more often to university marketing efforts, even if just to say they are not interested in the technology. Historically, companies in the high tech sector are more likely to be members of industrial affiliates programs, but are less likely to license inventions from universities. When OTL markets an invention, it always asks companies for feedback on the technology, including reasons why the company is not interested. This information is helpful for a technology transfer office to evaluate the commercial potential of the technology.

Patent Decisions

Patent decisions are based on the intelligence obtained in the foregoing evaluation and marketing phases, but they are rarely easy. Nevertheless, decisions must be made. So, it will boil down to one question: Does the invention have the potential to produce meaningful income? The answer to that question is important because it takes real money to cover the direct and indirect costs to file, prosecute, and maintain patents. Thus, one good practice to set some threshold, a revenue bar if you will, under which one would not seek patent protection. For example, if one could not expect to generate at least $100,000 over the life of a particular patent, then it may not make business sense to file a patent application. Typical patenting costs for a U.S. patent are about $25,000 at a minimum and can easily exceed twice that amount depending on the invention's complexity and prior art. Foreign patent costs will cost around $300,000 to maintain coverage in the key western economies. So, even with just U.S. patent coverage, a revenue bar of $100,000 is on the border of being a reasonable return on investment. While universities are not trying to maximize revenue, prudent business sense requires that some revenue threshold be established as a standard practice.

There are other factors that can be considered when deciding to file patent applications. One is the track record of the inventor. If an inventor has had successfully licensed inventions in the past, then it could be a reasonably safe bet to roll the dice, as it were, and file a patent application even if it's not possible to thoroughly evaluate the invention's licensing merits.

Another factor that could be used is assessing whether certain technological trends are relevant to the invention's adoption. One interesting example is the invention that formed the basis for Google. In 1996 when then Stanford graduate student Larry Page submitted his invention disclosure on his Hypertext

Search algorithm to Stanford's OTL, there were already several commercial search engines in existence. None of the companies with which Stanford OTL met expressed more than passing interest in the new search technology. Some outright did not see any value in it at all. There appeared to be a saturated market for search engines. Worse, none of the search engine companies were profitable. So, ostensibly a new search technology did not look very promising from a licensing perspective, at least at *that* time. This is where looking at trends can be important. Stanford OTL knew that the internet was still in its infancy and had great growth potential. In spite of the dearth of license interest, investing in a patent application on this new search technology was assessed to be a worthwhile risk. Worthwhile indeed, licensee Google showed what could be done with the Stanford technology and that patent investment eventually returned over $300 million to Stanford.

Filing foreign patent applications is costly as previously mentioned. One approach to minimize initial costs is to file a patent application through the Patent Cooperation Treaty (PCT) route. Filing a PCT application can delay the majority of foreign filing expenses for up to thirty months after a priority filing date is obtained in the U.S. Those thirty months can be used to further assess an invention's worth and to look for licensees. Also, by maintaining foreign filing rights, IP managers preserve the value of the invention for many transnational companies. At the end of the 30 months, if a licensee has not been found, then the decision to proceed with foreign applications would have to be based on very a confident assessment that makes the added risk justifiable.

Another important consideration whether one files patent applications in the U.S. or outside is the ability to enforce one's patent rights. The U.S. has good, strong patent laws and enforcement mechanisms. However, if the invention's commercial embodiment is difficult to detect, then enforcement becomes problematic. For example, say you have a new process for making a chemical. Nowhere in the end product that is sold is evidence that your process patent has been infringed. Without the ability to easily detect infringement makes having a patent possibly not worthwhile.

Outside the U.S., some countries have good patent and enforcement laws like the U.S. has. Others do not. Clearly, even if infringement detection is easy, one must give careful consideration before investing in patent applications in countries that have weak patent enforcement authorities. Furthermore, in countries that have strong enforcement authorities, the costs to sue a foreign company are not only considerable, given likely language and law differences, but the distraction to one's business could be quite significant.

There are many considerations that go into patent filing decisions that are done on a case-by-case basis. If a large portfolio of patents is to be maintained, then one should expect to need a significant budget to cover all the patent filing, prosecution and maintenance expenses. For example, Stanford's annual patent expenses exceed $6 million to maintain about 5,000 patents and patent applications. The management of all those patents also requires a substantial overhead. Because new inventions are being evaluated daily, a very rigorous reassessment must be done continually in order to keep the patent portfolio manageable.

License Negotiations

Once a company has expressed an interest in obtaining a license to a technology, OTL begins the negotiation process by providing the company with a proposed set of terms. The terms include all of the financial payments associated with the potential license and certain definitions of the scope of the license. Table 6.2 highlights many of the areas negotiated at the term sheet stage.

Once the basic terms are agreed on, Stanford uses its boilerplate agreements for each license or option agreement. These agreements can be viewed at http://otl.stanford.edu/industry/resources.html#documents and contain Stanford's standard definitions for the items listed above.

Although much of the agreement is negotiable, there are certain clauses to which Stanford rarely makes changes, such as warranties and indemnification. Other clauses may even be nonnegotiable. One is Stanford's retained rights, which states:

> Stanford retains the right, on behalf of itself and all other non-profit academic research institutions, to practice the Licensed Patent and use Technology for any non-profit purpose, including sponsored research and collaborations. Licensee agrees that, notwithstanding any other provision of this Agreement, it has no right to enforce the Licensed Patent against any such institution. Stanford and any such other institution has the right to publish any information included in the Technology or a Licensed Patent.

Industry is often concerned with this clause for many reasons. First, Stanford retains the right to use the Licensed Patent (and any associated technology developed at Stanford) for any non-profit purpose, which includes sponsored research and collaborations. Since these sponsored research and collaborations could include industrial partners, potential Licensees are concerned on many levels if they have a fully exclusive license to the technology. At the same time, Stanford is most concerned about maintaining its right to accomplish its education and research missions, which are far more important to the university than licensing inventions. Then Stanford also retains this same right for all other nonprofit academic institutions. However, even though sponsored research and collaborations with other companies are allowed, that does not allow those companies to practice the technology outside of those nonprofit academic institutions. In addition, any purpose other than nonprofit purposes is prohibited. By policy, Stanford does not make changes to its retained rights clause.

TABLE 6.2.   Areas Negotiated at Term Sheet Stage

Licensed Patent	A definition of the patents and/or patent applications that will be granted to the company licensing the technology ("Licensee").
Licensed Field of Use	The commercial area or application in which the Licensee can practice the Licensed Patent, for example, cancer therapeutics or wireless communications.
Licensed Product	Any product in the Licensed Field of Use that would infringe on a Licensed Patent.
Licensed Territory	The areas of the world the Licensee can practice using the Licensed Patent or sell Licensed Product.
Exclusivity Term	If the license is exclusive, how long the exclusivity lasts. Examples include: a) five years from date of first commercial sale and b) eight years from the effective date of the license agreement.
Diligence	For a nonexclusive license, the diligence provision states that Licensee will diligently develop the technology and, if it will be offered for sale, a date by which a first commercial sale must be made. For an exclusive license, yearly diligence milestones are included in the license, for example.
Upfront Payment	Also known as a license issue royalty, the amount negotiated that the Licensee pays to the party granting the license ("Licensor") upon executing the license agreement.
Annual Minimum	Also known as annual royalty or a maintenance payment, the annual minimum is the amount the Licensee pays to the Licensor each year, usually on the anniversary of the effective date of the agreement. This amount is a diligence payment and is often creditable toward earned royalty payments.
Earned Royalties	Also known as running royalties, the amount paid to the Licensor based on the sales of the Licensed Product.
Milestone Payments	When certain milestones toward developing a product are achieved, the Licensee will pay to the Licensor certain amounts. These payments recognize the increased value of the technology and often compensate for lower upfront and annual minimum payments.
Assignment Fee	Licensor will allow the Licensee to assign the license agreement in certain instances (such as a sale of the business), and in consideration of this permission, the Licensee will pay the Licensor a certain fee upon the assignment.
Sublicensing	If the Licensor allows the Licensee to sublicense (which is usually the case for exclusive licenses, but not for nonexclusive licenses), the Licensee pays the Licensor certain amounts for granting the sublicenses. This amount can be percentages of the amount the Licensee receives or set amounts.

| Patent Expenses | In the case of exclusive licenses, the Licensee will pay for the patent expenses associated with the Licensed Patents. |
| Patent Enforcement | In the case of exclusive licenses, Licensee will have some rights to enforce the patent against potential infringers. he section in the license describes how Licensor and Licensee decide who will pursue the potential infringers and how any recoveries will be split. |

Another of the major concerns companies have in the retained rights clause is that the university and other institutions can freely publish on the technology. Again, Stanford is upholding its mission by maintaining an open environment and protecting its pursuit of academic research and education. Faculty and student credentials include their peer-reviewed articles, and Stanford will not have licensing activities inhibit any research, education, or career activities of its researchers and students. In industry, prohibitions on publications are common in order to protect trade secrets, intellectual property rights, and competitive knowledge, but when dealing with an academic institution, the industrial partner should recognize the importance of preserving academic freedom.

Other areas in the license agreement that are different from agreements between companies are the indemnification, liability, and warranties clauses. By policy, most academic institutions will not generally provide warranties and require indemnification by the Licensee. Stanford can warrant that it can enter the license relationship and that it holds the Licensed Patent, but cannot guarantee anything regarding the patenting, the development, or the commercial viability of the technology itself. The university also requires that a company indemnify Stanford for any claims that arise as the result of the license agreement and that Stanford is not liable for any damages arising from the use of the rights granted under the agreement. Licensing is not part of the Stanford mission, and therefore the university does not want the licensing to endanger the mission through lawsuits that could strain the university's financial health.

Throughout the negotiation, Stanford's OTL concentrates on a main goal—establishing a good home for the technology. An exclusive license agreement includes yearly diligence clauses in order to ensure the Licensed Product is faithfully developed and not shelved in deference to other technologies or for anti-competitive motives. Moving a new technology from the lab to the marketplace is no small feat, and if a Licensee misses or is going to miss a diligence milestone (which would not be unheard of), Stanford would like to know as soon as possible so that it can discuss the issue with the company and decide whether to: 1) grant an extension for that diligence time point; 2) revise or negotiate new diligence milestones and time points; or 3) terminate the license. Stanford chooses termination only as the last resort, but it does include a right to terminate under the license agreement if the Licensee does not achieve its diligence milestones.

Maintaining Relationships

Universities seek licensees to transfer technology from the lab to the marketplace and to generate unrestricted income to support their research and education missions. Companies want to sell products to make profits for their shareholders. So, it seems that a very synergistic partnership can be made between universities and companies, whereby companies use university technology to make profits. It seems too good to be true! The problem historically is that universities have been seen as ivory towers and companies have been seen as greedy capitalists. Fortunately, those perceptions have faded, and universities and companies are starting to realize that there are benefits to working together. But the key to university/company relationships lies in understanding and communication. Companies cannot and should not expect universities to behave like other companies. Likewise, universities need to understand that businesses operate in a time-constrained competitive environment.

Many (if not most) partnerships between corporations end up in ruins. Part of the reason is the changing nature of business: business objectives change, corporate managers change, markets change. Universities, on the other hand, don't change much, if at all. They can make good partners because they are predictable. But, if in the rare case a dispute arises between a company and a university, say over a license provision, it is very likely that the parties could work out a settlement without having to resort to hired guns.

EXPECTATIONS WHEN WORKING WITH UNIVERSITIES

What should industry expect from a negotiation and license with a university technology transfer office? First, many university technology transfer offices do not break even on their operating expenses and are usually understaffed. Time and patience may be needed. However, most university technology transfer professionals sincerely enjoy their work and value the relationships they have with industry. In fact, many came from industry themselves and understand the company's perspective.

As mentioned previously, there are certain nonnegotiable clauses in university license agreements, but in other areas universities are often very flexible. For financial terms of an agreement, all terms are a balancing act. For example, if the company wants a lower upfront because of its financial status or its belief that the technology is very risky, that might be balanced by higher payments at future dates. Diligence provisions are often taken from a company's proposed development plan for the technology, and Stanford encourages adding a bit of extra time for the milestones to allow for unexpected events.

Stanford's template agreements are all in plain language and are crafted with the goal that any future person reviewing the agreement will be able to understand the intent of the clauses. Companies working with attorneys that have previously negotiated with the university may save valuable resources for the companies since the attorneys are already educated about the university's policies and license agreement provisions.

In Stanford's case, when a technology is licensed, only the technology as it stands at that date is licensed. Any improvements are not included and would need to be licensed under a separate agreement. Stanford does not include improvements because those improvements may not be reflected in the licensing payment amounts, and they could be subject to other third-party obligations. However, the licensee may be the best organization to license and develop the improvement, especially if it is dominated by the previously licensed technology; therefore, Stanford may grant rights to the improvement technology under a new license agreement to the company.

Even if the company is willing to grant Stanford equity in the company, Stanford still expects some sort of cash royalty up front upon signing the agreement. Other universities are willing to accept only equity up front, or are even willing to negotiate all equity licenses. Companies should expect to find varying policies regarding types of payments accepted by different universities. In Stanford's case, Stanford accepts equity in lieu of higher upfront payments, and possibly to grant some lower payments than it would otherwise expect down the road. However, Stanford still considers equity very risky and does not expect to make much money, if any, from the equity. Of the more than 160 companies in which Stanford has taken equity, three have brought in over $1 million to Stanford from the sale of equity—Abrizio, Amati Communications, and Google.

Stanford is looking for companies that will diligently develop the technologies created at Stanford and will enter into a fair deal with the university. It seeks to establish a communicative relationship with potential licensees, which is necessary for a connection that may last for twenty or more years. A good rapport helps the parties overcome any hurdles that develop in the company, the license, or the technology's development.

CURRENT CHALLENGES IN TECHNOLOGY TRANSFER

Universities face so many challenges when operating an IP management program that it is not possible to list all of them in a short chapter. Clearly, one of the most difficult ones is that this business is so darn difficult—licensing intellectual property is inherently tough. Add to that the prospect that an invention might not be commercially viable for ten or so years, and then you get the picture that a lot of patience is needed.

Another difficult challenge for universities is patent enforcement. A patent holder without the willingness to sue an infringer is about as threatening as a paper tiger. How is a university expected to be viewed as a good corporate partner on one hand, while on the other hand it is willing to sue a company for patent infringement? Will universities be viewed as mercenaries? Probably not, and for a few reasons: 1) Universities seldom sue for patent infringement; 2) Universities should go to extreme efforts to avoid litigation by attempting to work out settlements; and 3) the government has been supportive of universities taking legal action against infringers because the government (in most cases) paid for the invention and encouraged the university to patent the idea.

In negotiating with universities and institutions, companies will notice that the policies regarding intellectual property differ between universities. For example, in the U.S., universities rarely allow the sponsors of research to take title to inventions created through the research they sponsor. In Europe, the grant of title for sponsored research varies from country to country, and sometimes from institution to institution, though the general trend appears to be moving toward not granting title to a sponsoring company. This is a challenge for both the universities and the companies because the companies do not know what to expect from the universities, and the universities have to educate companies about their policies.

Intellectual property policies at universities can change, as they recently have in many countries. For example, before 2002, professors at German universities owned the titles to the inventions created at their universities. Since 2002, the universities own the titles to the inventions, but it takes time to adjust the faculty's mentality toward the university ownership and to set up well-operating technology transfer systems. In other countries, university inventors still own the products of all their research. Companies need to know the ownership laws for the countries and universities with which they work.

When licensing an invention, companies also find that institutions have different policies regarding improvement or other future related inventions. As mentioned previously, improvements (including continuation-in-part patent applications) are not included in license agreements with Stanford, as Stanford only licenses technologies it is aware of at the time of the license. Stanford does not license unknown improvements or other future technologies, as the value of those technologies is not known and they may be subject to other obligations, such as sponsorship rights. Other universities may include future improvements or related technologies if their policies and sponsorship obligations allow them to do so.

Countries also have differing laws regarding who may license and practice patent rights if there is co-ownership. If a patent is co-owned by U.S. entities, each co-owner may practice and license the technology as it wishes. However, a co-owner may not grant a fully exclusive license to the patent without the consent of the other co-owner. In Japan and Germany, as well as in other countries, neither co-owner may grant any licenses to the patent without the consent of the other co-owner.

Laws regarding inventorship also vary from country to country. In the United States, the true inventors must be noted on a patent; otherwise, it is possible for the patent to be invalidated. The U.S. grants patents only to the inventors who create an invention, which is known as a "first-to-invent" system. Therefore, noting correct inventorship is important. In other countries where a "first-to-file" system is in place, true inventorship is not as important because the person or entity who first files the patent application for an invention will be granted the patent if the technology is patentable. Since the importance of actual inventorship is diminished, in some countries there is a practice of adding inventors from companies to university patents as thanks for a company's support. The company therefore has rights to the patent. As noted previously, if this occurs in Japan, the university does not have the right to otherwise license the technology since it is co-owned with another entity, unless the other entity gives its permission.

These differences in intellectual property management policies and laws provide challenges to universities, institutions, and companies because they may not know what to expect when dealing with a new party. Negotiations may take much longer or break down because the parties' expectations were unfulfilled. When entering into negotiations, each party should be or become acquainted the other's intellectual property policies to avoid surprises or questions on ownership.

Another key challenge for universities is attracting interest in their inventions from high-tech companies (i.e., electronics, communications, and the like). Part of the reason has to do with the N-I-H syndrome, but the other part has to do with the nature of high-tech companies' product life cycles and typical cross-licensing arrangements (that don't benefit universities). As global business gets more competitive and profits get further pinched, high-tech companies might be well rewarded if they can take advantage of university-based inventions. Granted, a challenge is presented by the usual fact that university inventions are typically early stage. But there will be those inventions that came about five years ago that might be ripe for the picking. The best thing about the deal? It doesn't cost anything to look!

CONCLUSION

Despite the current and past challenges, university technology transfer activities are flourishing. An increasing number of licenses and partnerships with industry are forged each year. Countries worldwide are developing their own policies and technology transfer organizations.

Thousands of university researchers worldwide are creating inventions daily that will change our lives. Stanford now receives one invention disclosure a day. Perhaps the invention received tomorrow will be an HIV vaccine or a faster semiconductor chip. However, universities are not structured to fully develop a medical product or to manufacture electronics for wide-scale use, so without assistance from industry, most innovations would languish in the laboratories.

This is a two-way street. Universities and other nonprofit institutions provide companies opportunities beyond their internal research labs. Research is expensive, from the people to the time to the labs, including supplies and equipment. Nonprofit research institutions are on the cutting edge of research, paving the roads to the future. Partnering between universities and companies provides each with means to accomplish their missions, and proven mechanisms are in place to bridge the cultural gap inherent between these two different types of entities.

NOTES

1. National Science Foundation/Division of Science Resources Studies, "Survey of Research and Development Expenditures at Universities and Colleges, Fiscal Year 2002," Table 32. Data from top 100 universities.

2. 2005 R&D Scoreboard: http://www.innovation.gov.uk/rd_scoreboard/

3. Association of University Technology Managers Licensing Survey (http://www. autm.net/surveys/). These products provide new taxable revenue sources to the Treasury and may help taxpayers in the form of new medicines or medical devices.

4. Association of University Technology Managers Licensing Survey (http://www. autm.net/surveys/). Data varies according to number of survey respondents. 2003 Research Budget data includes 188 university respondents.

5. Stanford University Research Policy Handbook, Document 10.5, November 1995. http://www.stanford.edu/dept/DoR/rph/10-5.html.

6. U.S. Government Accounting Office (GAO) Report to Congressional Committees entitled "Technology Transfer, Administration of the Bayh-Dole Act by Research Universities," dated May 7, 1998.

7. Under Stanford royalty sharing policy, cash royalties are split as follows: 15 percent is taken off the top for OTL's budget, and then any out-of-pocket expenses are deducted. The net amount (usually 85 percent) is split into thirds: one-third to the inventors, one-third to the department, and one-third to the school.

8. A waiver allowing the substantial manufacture in another country may be obtained from the U.S. government in some cases.

9. H.R.6933, Public Law 96-517, passed December 12, 1980, and H.R.6163, Public Law 98-620, passed November 8, 1984.

10. If the invention has external sponsorship, the university may have an obligation regarding the filing of patent applications.

11. U.S. Code, Title 35, Part II, Chapter 18, Section 200.

12. http://stanfordtech.stanford.edu/technology.

APPENDIX

Technology Evaluation Worksheet

Invention Title:_____

Inventors:_____

Docket #_____ Date:_____ Associate:_____

Commercial Potential	(+)	(0)	(-)*
• Ability to define product (i.e. what is for sale?)	—	—	—
• Perceived need	—	—	—
• Identity of end user	—	—	—
• Market size	—	—	—
• Maturity of market	—	—	—
• Competitive advantage/product differentiation	—	—	—
• Prospective licensee(s) identified	—	—	—
• Liability considerations	—	—	—
• Predisposition of industry to licensing	—	—	—

Inventor Profile			
• Cooperative/will serve as champion	—	—	—
• Industry contacts	—	—	—
• Realistic expectations	—	—	—
• Success with previous disclosures	—	—	—
• Credibility/recognition in field	—	—	—
• Research funding and direction	—	—	—
• Conflicting obligations	—	—	—

Scientific/Technical Merits			
• Invention is adequately defined in disclosure	—	—	—
• Supporting data are available	—	—	—
• Utility shown (i.e., solves a problem)	—	—	—
• Core technology versus improvement	—	—	—
• Features of invention versus limitations	—	—	—

Proprietary Position/Patentability Issues			
• Patentability of invention (new, useful, non-obvious)	—	—	—
• Breadth and strength of claims	—	—	—
• Freedom to practice (i.e., other dominant patents?)	—	—	—
• Possibility of reverse engineering	—	—	—
• Ability to detect infringement	—	—	—
• Ability to withstand litigation	—	—	—
• Known prior art exists	—	—	—

Stage of Development	(+)	(0)	(-)*
• Only a concept	—	—	—
• Reduced to practice/prototype available	—	—	—
• Manufacturing feasibility (facilities, equipment, etc.)	—	—	—
• Clinical data available	—	—	—
• Inventor cooperation required	—	—	—

Avenues of Commercialization			
• License to established company	—	—	—
• License to start-up company	—	—	—
• Access to venture capital	—	—	—
• Package with other technologies	—	—	—
• License to inventor	—	—	—

Financial Analysis			
• Cost of patenting	—	—	—
• Financial support from licensees	—	—	—
• Possibility of sponsored research	—	—	—
• Anticipated license/royalty income	—	—	—

Total:	—	—	—

Comments:_____

_____

_____

_____

_____

_____

_____

Recommendation: _____Needs further research

_____Return to Sponsor/Inventor(s)

_____Patent application

_____Market without patent

* **Ranking Codes:** (-) unfavorable   (+) favorable   (0) neutral

7

Social Innovation

KRISTI YUTHAS

I n its purest form, social innovation is the use of new and creative ideas to generate environmental or societal benefits. Organizations in both the commercial and social sectors have realized enormous returns from social innovation. Traditional views of business innovation typically measure the success of an idea in purely commercial or economic terms. Social innovation, on the other hand, expands the definition of success to include a broad range of environmental and societal outcomes. From the perspective of social innovation, success can be measured by improvements in the workplace, the community, or the natural environment—it includes a myriad of social outcomes valued by an organization or its stakeholders. Whereas commercial innovation is often stifled at the first hint of commercial infeasibility, social innovators are often free to pursue the desired benefits first, and tackle the economic challenges later. In a world in which improvements in organizational social performance are valued and often expected, economic and social outcomes are increasingly intertwined, and social innovations often result in long-term economic benefits.

A plan for innovation that specifically includes social objectives can produce impressive benefits for both commercial and social organizations. A commercial organization that encourages social innovation may discover a revitalized sense of passion and creativity among its employees. A newfound energy is often the result of working toward social goals, and the organization can harness that energy and use it to enhance the overall effectiveness of the organization. Social organizations, on the other hand, can apply the tools and discipline of business innovation to achieve breakthroughs in social effectiveness. They can create systematic ways to capture the innovative spirit, which drives the organization's mission and turns it into productivity and effectiveness gains. This chapter provides a starting point for thinking about social innovation. First, it discusses the evolving concept of social innovation and the types of organizations that benefit from it. Then it describes specific methods that organizations can employ to engage in and create value through social innovation.

Organizations today have many reasons to think seriously about their social impacts. Some organizations are reactive—seeking social improvement as a response to stakeholder pressure. Others take a proactive approach, and view management of social performance as an opportunity. These organizations recognize that ultimately, in addition to improving social performance, social innovation can lead to significant improvements in operational and financial performance as well (Berman et al. 1999). A number of current trends provide incentives for engagement in social innovation:

- stakeholders becoming more knowledgeable and engaged;
- corporate power and accountability are increasing;
- opportunities for generating social value are more widely available; and
- social performance is an important driver of economic performance.

Organizational activity can have a broad range of social consequences. Among the areas of greatest interest to stakeholders are product safety, labor rights, human rights, environmental impact, and community involvement. An organization's impact on society in these areas can be quite dramatic, and stakeholders are becoming increasingly aware of the social effects of organizational activity (Bansal and Roth 2000).

Businesses are also recognizing that along with increased stakeholder awareness come increased influence and power. Stakeholders continuously monitor corporate activity and demand that organizations deliver not only good economic performance, but good social and environmental performance as well—these demands are well justified (Sharma 2000).

Global corporate activity eases the movement of capital and ideas across political boundaries. This deterritorialization reduces the power of states and increases the burden on organizations to manage social outcomes. As responsibility for social outcomes grows, the risk of ignoring social concerns also increases. When corporations ignore social outcomes, they do so to their peril. When expectations for social performance are violated, market discipline for social irresponsibility can be severe and companies can quickly lose relational capital that took years to build (Schneitz and Epstein 2005). Nike, for example, remains a target of intense scrutiny after low wages and poor working conditions in the factories of its Indonesian suppliers were made public over a decade ago.

But social innovation is much more than a responsibility—it is also an area of enormous opportunity and potential. Nike has learned this lesson as well. In reengineering its global supply chain to control and improve factory conditions, Nike production managers identified and shared best practices and developed metrics systems that not only improved factory conditions but also streamlined the supply chain, reduced waste, and improved economic performance.

This story is by no means unique. Companies seeking to improve environmental performance through environmental management systems commonly report a host of unanticipated benefits (Christman 2000). In addition to adding business value through cost reduction and improved control, companies find that as they create continual improvements in environmental outcomes, they gain spillover effects that lead to both incremental and radical innovations in other areas of business activity. Drivers of these innovations include institutional knowledge capture, increased openness toward change, and enhanced employee engagement and commitment.

SOCIAL INNOVATION

Social innovation is the generation and implementation of ideas that lead to social value. Social innovation in organizations has not been thoroughly explored or researched, and as yet, no single definition of the concept has been widely acknowledged. Social value or well-being is a complex, multidimensional construct. The notion of desirable social objectives or standards can vary widely across culture and time, and perceptions can differ greatly among the many groups holding a stake in those outcomes—employees, customers, suppliers, investors, regulators, and community members, among others.

As defined from a business perspective, social value generally encompasses both environmental and societal dimensions. Key environmental outcomes include the use of materials, water, and energy, the emission of wastes, the life cycle and use of products and services, and impact of organizational activity on health and biodiversity. Societal outcomes include human rights issues such as discrimination, forced and child labor, and health and security in the workplace. They also include issues of product safety, privacy, competition, and corporate governance, as well as poverty, peace, political influence, and other community impacts.

Understanding the complex web of social and economic consequences of organizational activity is exceptionally difficult now, and the challenges continue to mount as worldwide economic, political, and social systems intertwine (Steyaert and Katz 2004). More than ever before, effectively managing the consequences of organizational activity will require sustained and systematic social innovation.

For individual organizations, social innovation is increasingly important. For many, it will play a key role in long-term organizational success (Paine 2002). It is widely recognized today that competitive advantage in any organization, regardless of the source of this advantage, is temporary. One well-known perspective holds that advantage is built upon firm-specific resources that are rare, valuable, and difficult to imitate or substitute (Wernerfelt 1984; Barney 1991). Yet to the extent that such resources generate value for the firms possessing them, these resources will be the target of imitation efforts. Competitors rapidly move in to high-potential markets and mimic the processes and products of successful firms. In dynamic environments, a key factor in developing and sustaining competitive advantage is therefore the organization's capacity to learn and innovate (Teece, Pisano, and Shuen 1997).

Like other forms of innovation, social innovation seeks the successful exploitation of new ideas. The ideas that underlie innovation are novel—they can be new ideas developed through formal research or existing ideas applied in new ways or in new contexts. They can represent small incremental changes or radical disruptions to the status quo. Innovation requires creativity in generating new ideas as well as creativity in putting these ideas into practice. Innovators must therefore possess both the ability to generate promising ideas and the knowledge and skill to develop these ideas into valuable firm resources such as policies, processes, or products.

In traditional thinking about business innovation, success is envisioned in commercial terms. An innovation is deemed successful when an idea can ultimately be used to generate wealth for the firm. Although the processes required for social innovation are analogous to those required for business innovation, their objectives differ—business innovation seeks primarily to generate economic value, while social innovation pursues social benefit by generating environmental, societal, and/or economic value. The two forms of innovation also differ dramatically in their perceived importance and in the resources available for their pursuit. While commercial innovation has always been a requirement for the long-term viability of organizations, social innovation has, until recently, been addressed most heavily by institutions and systems outside of the market. And the organizations and institutions that were engaged in social innovation often lacked the structures and systems to optimize the outcomes of their efforts. Today, organizations of many types are beginning to recognize the importance and value of social innovation.

ORGANIZATIONAL FORM

Social innovation can be pursued through many different organizational forms. These forms fall across a spectrum that ranges from traditional commercial businesses, such as Exxon or General Motors, to social sector organizations such as the Salvation Army or the American Cancer Society. The most important distinction between these two types of organizations is in the objectives they pursue. The primary objective for a commercial organization is normally generation of financial benefits for owners and other stakeholders. The primary objective for a social organization is the maintenance and improvement of social conditions. Hybrid organizations such as Ben and Jerry's and The Body Shop simultaneously seek financial and social goals—often by pursuing financial gains through socially-beneficial methods. Hybrids can be stand-alone organizations, or they can be partnerships between social and commercial organizations.

The lines between social and commercial organizations are increasingly blurred. The long-term survival of many social organizations demands economic viability while, at the same time, corporations are increasingly accountable for social conditions in the communities in which they operate (Waddock and Graves 1997). Actions directed toward social value are taken by organizations of all forms, including social sector organizations, traditional commercial enterprises, and hybrids that combine elements of both through traditional forms as well as through partnerships, alliances, and joint ventures.

Table 7.1 highlights basic characteristics of these forms as they relate to social innovation.

TABLE 7.1. Social Innovation and Organizational Form

	Social Organizations	Hybrid Organizations	Commercial Organizations
Mission	Primarily social mission	Combined commercial and social mission	Primarily commercial mission
Business model	Noncommercial and commercial activity support pursuit of social outcomes	Commercial activity supports pursuit of both social and commercial outcomes	Social activity supports pursuit of commercial outcomes
Examples of organizational forms	• Donation-driven not-for-profits and non-governmental organizations • Revenue-driven not-for-profit and non-governmental organizations	• For-profit organizations with social agendas • Joint ventures, partnerships, and alliances between social and commercial organizations	• Dedicated units within commercial organizations (e.g., corporate social responsibility, environmental health and safety, and public relations) • Research and development unit
Site for social innovation	Embedded throughout organization	Embedded throughout organization	Isolated in units responsible for social impact or innovation

Social Organizations

Social organizations are organizations that adopt social missions and work to provide social benefits. These organizations are constantly engaged in social innovation, as they work to fill needs unmet by traditional market mechanisms (Seelos and Mair 2004). Social organizations have traditionally engaged in noncommercial activity supported by donations and grants. Today, many social organizations also engage in revenue-seeking activities geared directly toward achievement of social goals or as a sideline source of funding for social operations. The forms most commonly represented in discussions of social innovation are not-for-profit organizations (NPO) and nongovernmental organizations (NGO). NPOs are generally charities, service organizations, or foundations that are formed to pursue a specific social purpose, such as providing products or services to improve social conditions. NGOs are typically not-for-profit organizations formed to pursue development or advocacy goals by influencing institutional decision-making and policy formation.

Social organizations play an increasingly important role in the global economy. In the U.S. alone, there are approximately 1.5 million chartered not-for-profit organizations. The number of social-sector organizations and the number of people employed by those organizations are growing rapidly. With the movement toward privatization of public services in many regions combined with pressure for greater corporate involvement in social and community issues, the nature of the social sector is changing rapidly.

Social innovation in these firms is embedded throughout all aspects of the organization, and the structure and processes of these firms are aligned with the social mission and strategy. In many cases, the very existence of these organizations is a result of innovation in a basic business model or value proposition. Social organizations are commonly formed to respond to a perceived gap in the marketplace for satisfaction of social needs that are left unmet by for-profit and governmental organizations (Shore 1995). To fill the gaps, social organizations develop innovative ways to maintain or improve social conditions, or to prevent social problems through proactive risk management (Nicholls 2006).

In addition to innovations in the services provided and markets served, social organizations are innovators in a variety of other critical arenas, such as in their communication systems, the mechanisms used to deliver the services, and public relations activities. In addition, social organizations must continually innovate their approaches to fundraising and development of financial self-sufficiency (Porter and Kramer 1999). While some NPOs are funded entirely through grants and charitable donations, these organizations increasingly turn to revenue-seeking activities in order to establish a continuing source of funding. The term *social enterprise* is often used to refer to not-for-profit organizations that sell goods or services as a means through which to generate revenue to accomplish social aims. In addition to sales of goods and services, these enterprises have pursued revenues through activities such as capital asset and property rental, use of patents and copyrights, branding and licensing agreements, and investments (Emerson and Twersky 1996).

Hybrid Organizations

Hybrid organizations are organizations that pursue combined commercial and social missions through commercial operations. They can take the traditional form of a for-profit organization that combines a social agenda with economic goals, or they can take a variety of less traditional forms, such as joint ventures, partnerships, or alliances formed between social and commercial organizations. As with social organizations, social concerns in hybrids are integral to the mission and culture, and social innovation is embedded throughout the organization.

Hybrid organizations can possess strong social missions supported by commercial operations or strong commercial missions accompanied by social agendas and values, or they can fall somewhere in between. Wherever they lie on the spectrum, these organizations pursue activities that integrate social and commercial goals. In these organizations, social and commercial interests are tightly intertwined, and economic and social success go hand in hand.

Hybrid organizations that operate as standalone for-profit organizations generally pursue their social agendas in one of two basic ways. They can meet social needs directly through the products or services they provide and the markets they serve, or they can provide traditional products and services, but produce or distribute them using socially beneficial methods or processes.

Organizational forms that represent collaborations between business and social organizations represent innovative means for pursuing social goals. Many corporations have philanthropic relationships with social organizations, contributing directly through donations, sponsorships of events, or volunteer activities, or indirectly through foundations and similar forms; however, two-way collaboration between social and commercial organizations is becoming more common. In these relationships, both organizations accrue strategic benefits through collaboration. These organizations can take the form of independent joint ventures, or they can take the form of partnerships, alliances, or other forms of collaboration (Austin 2000).

Social/commercial collaborations take advantage of strategic assets of both firms, such as the powerful missions, strong public image, or community relationships of the social organization, or the organizational expertise, facilities, or distribution channels of the commercial organization (Timmons 1994). Collaboration can include a broad range of activities, such as cause-related marketing, in which a portion of the proceeds from a campaign goes to the social organization, licensing agreements in which a commercial organization or its products are endorsed by a social organization, or contracting relationship in which services such as health care or education are provided by the social organization.

Commercial Organizations

Commercial organizations pursue primarily economic missions. These organizations engage in social activity and innovation because it supports them in their pursuit of commercial outcomes. In these organizations, specific organizational units are generally responsible for management of the corporation's social impact. Corporations may locate social pursuits in dedicated corporate social responsibility (CSR) units, responsible for managing relationships and impact relating to a broad range of corporate stakeholders. Or they may focus more heavily on legal or public image aspects of their social outcomes, locating social activity in environmental, health, and safety (EH&S) units or public relations (PR) units. Companies aggressively pursuing performance through social innovation may locate responsibility for social outcomes in a corporate research and development (R&D) unit.

Commercial organizations engaged in social innovation may pursue this activity as a direct means of enhancing the business's value proposition, for example, through cost reduction, liability reduction, or increased access to desirable markets. As discussed earlier, managing social impacts is imperative for large global organizations and is becoming increasingly important for organizations of all types. Social innovation has therefore become a priority for commercial organizations (Mirvis and Googins 2004).

Strengths and Opportunities

While both social and commercial organizations are engaged in social innovation, the strengths they bring to their innovative activities differ in substantial ways, as do their opportunities for improving social outcomes through innovation. Historically, social organizations are strong in mission and culture, but are less so in the rigors of value chain activities; traditionally, commercial organizations are strong in business disciplines, but lack the motivation and cohesive culture that derives from shared values. Table 7.2 provides a brief summary of the strengths and opportunities of typical social and commercial organizations.

TABLE 7.2. Strengths and Opportunities for Social Innovation

	Social Organizations	Commercial Organizations
Strengths	Mission and value proposition Organizational commitment Leveraging scarce resources	Management expertise Value chain practices Operational efficiency
Opportunities	Business discipline Business innovation	Passion and commitment Social improvement

Social organizations exhibit strength through their basic mission and value propositions. These organizations often possess clear and powerful missions focused toward real and significant social needs. As a result, they often develop strong organizational cultures surrounding social values, and attract employees committed to the organization's effectiveness in pursuing them (Brinkerhoff 1994). In general, these organizations operate in environments in which financial and other key resources are scarce, so they develop creative means to use sparse resources to maximum organizational advantage (Dees, Emerson, and Economy 2001).

Commercial organizations generally operate in competitive industries. Competitors, operating under similar pressures or possessing desirable competencies, can serve as role models and sources of learning and evaluative comparison. The results are opportunities for development of management experience and expertise, development of effective business practices across the value chain and in supporting activities, and a focus on operational efficiency that aligns a company's costs with that of its competitors.

Each type of organization can benefit by developing the strengths possessed by its counterparts. Social organizations can benefit from more rigorous discipline in traditional business functions and by adopting standard tools of business innovation. These tools provide the mechanisms needed for continual improvement and innovation in both the value proposition and the technology through which social value is generated.

Commercial organizations can benefit from more fully accepting social values and objectives. They can benefit from the passion and commitment commonly associated with pursuit of worthwhile social objectives. Additionally, they can profit from efforts toward social innovation that provide both social benefits and carryover benefits that lead to innovations resulting in commercial benefits.

MANAGING FOR SOCIAL INNOVATION

Like other forms of business innovation, social innovation can and must be managed for maximum organizational effectiveness. Davila, Epstein, and Shelton (2006) provide a model for business innovation that defines innovation and discusses how it can most effectively be managed. The authors argue that companies can innovate in two basic areas: technologies and business models. Technological innovations include improvements in the product or service provided, in the processes used for production, and in enabling technologies such as information systems. Business model innovations include changes in the customers targeted, the supply chain, or in the basic value proposition of what is sold and delivered to the market.

In social organizations, as in commercial organizations, innovation can be pursued through technology or business model change. Much of the current wisdom about innovation can be applied equally to social and commercial organizations. However, there are some key differences in the manner in which social and other innovations are likely to be pursued. For example, unlike many commercial organizations, a key area targeted for innovation in the social sector is likely to be financing activity—an important enabling technology needed to support execution of any business model in that sector. As noted above, social firms are also likely to pursue major changes in the value proposition, as they pursue revenue-generating activities or partnerships to augment traditional financing sources.

Likewise, in commercial organizations, social innovation can be pursued through any aspect of technology or business model change. However, for commercial organizations, business model innovations are associated with pursuit of social value. When firms make changes that significantly improve social outcomes, they simultaneously change the basic value proposition of the enterprise.

Davila, Epstein, and Shelton (2006) identify general rules for innovation and explore how the rules can be applied by using standard management tools such as strategy, structure, and process. The following discussion uses many of the same elements and concepts, and modifies them for a social innovation context. Table 7.3 highlights a variety of means through which social and commercial organizations can more effectively engage in social innovation through effective use of standard management tools. The analysis dichotomizes social and commercial organizations, although, as discussed above, all organizations are hybrids in that they can benefit from commercial business discipline applied in pursuit of social outcomes.

Leadership

Organizational leadership is the first, and perhaps most important, tool in promoting social innovation. Social organizations often have strong leaders who are heavily and personally committed to pursuing the organizational mission. Because social organizations generally operate in an environment of very restrictive resource constraints, leaders in these firms are forced to be flexible and innovative to meet normal business demands. Nonetheless, social firms can benefit from systematic efforts to lead innovation within their organizations.

To enhance innovation, social organizations need leaders who are passionate about both the basic value of continuous innovation as well as the importance of developing business practices that can effectively harness that innovation. Although the need to operate with minimal resources may encourage innovation, the existence of heavy demand for an organization's services can work against it, as building capacity to do more of the current activity might be the center of focus. Even without market pressures demanding innovations in the product or how it is produced, social organizations always face competition for funding. Competition becomes more severe as the rapid

TABLE 7.3. Tools for Managing Social Innovation

	Social Organizations	Commercial Organizations
Leaders	-understand commercial best practices	-identify social vision that complements commercial goals
	-recognize importance of innovation	-understand social opportunities and risks
	-build structures and process to manage innovation	-champion integration of social and commercial
Culture	-recognize need for social innovation	-integrate social ideals with values and identity
	-value commercial management practices	-open to exploration of social innovations and outcomes
	-open to commercial innovation	
Structure	-flattened structure	-greater integration between respect for social outcomes for others
	-cross-boundary communication mechanism	-funding/support mechanisms for social innovation
	-organization units charged with generating innovation	-structure for stakeholder engagement
Innovation strategy	-develop an innovation strategy	-incorporate social interests into innovation strategy
	-balance between idea generation and value capture	-focus on commercial benefits of social innovation
Structure	-create organizational units charged with innovation	-integrate social interests into innovation structure
	-create internal market for innovation	-develop structures for stakeholder engagement
Process	-formal business planning	-social opportunity and risk identification
	-enterprise risk management	
Performance measurement and rewards	-develop innovation metrics	-incorporate social objectives into innovation metrics
	-monitor financial performance	-monitor social performance
	-reward innovation	-reward social innovation and performance

growth in social organizations and tighter margins for commercial enterprises further restrict access to resources.

Leaders in social organizations must therefore recognize that the ability to accomplish a social mission can be greatly enhanced by pursuing continual improvement and innovation in all aspects of business practice. Effective leadership for innovation requires considerable knowledge of best practices in the tool of general management as well as the recognition that staying at the forefront of management practice requires continual innovation. More and more, social organizations are pressured by their funding sources to demonstrate not only the quality of social outcomes, but also the efficiency with which these outcomes were generated. Leaders must therefore take responsibility for building structures and processes to manage innovation throughout all areas of organizational activity. Examples of structures and processes that support innovation are provided in the following discussion.

Commercial organizations require different knowledge and capabilities from leaders as they pursue social innovation. First, leaders must develop a meaningful social vision that is directed toward important social considerations and is compatible with and complementary to the firm's commercial interests. To develop such a vision, commercial leaders need thorough knowledge of the social interests of key stakeholders and an understanding of how these interests can be translated into opportunities for the firm.

Interaction with stakeholders also provides a rich source for understanding organizational opportunities for social improvement and for identifying a broad range of environmental and social risks that might impact accomplishment of commercial objectives. Leaders must regularly assess these opportunities and risks facing the firm, and should develop plans to address them in a manner that is sensitive to the social and commercial strengths and weaknesses of the firm.

Perhaps most importantly, leaders must serve as champions—combining a passion for social objectives with the ability to convey the importance of these values throughout the firm. They must develop and communicate a vision of desirable social values and objectives that is compatible with and supportive of the firm's commercial mission and values. They should have the capacity to help others in the firm recognize and internalize social values and embed them into the everyday practices and decisions taking place throughout the firm. In situations within the firm when social pursuits are viewed as a threat to commercial effectiveness, leaders must be able to demonstrate how the organization gains commercial benefits through its socially oriented endeavors.

Culture

Social organizations tend to develop organizational cultures that embrace the organization's social mission and values. Employees throughout these organizations often align themselves more strongly with organizational values than do their counterparts in commercial organizations. Although these cohesive cultures are effective in motivating action aligned with the organizational objectives, the resulting homogeneity may have the unanticipated consequence of slowing innovation. Employees in these organizations may be less flexible regarding any innovation that risks modifying the organizational direction or ideals. But, like commercial organizations, continued effectiveness in accomplishing goals requires development of a culture that values and encourages innovation by systematically encouraging and rewarding innovative endeavors.

Efforts toward innovation must be balanced with and supported by business discipline. In social organizations, it can be difficult to engender a culture that embraces the value of tools and objectives associated with commercial organizations. Members of social organizations may view social and commercial values to be antithetical to one another, and may resist cultural changes that move the firm away from a singular focus on social objectives. Nonetheless, social organizations seeking innovation need to overcome this resistance and view efficiency and productivity as important contributors to social goals.

Social organizations can also benefit from cultures that are open to consideration of other forms of commercialization. This allows the organization the option to take advantage of promising opportunities for revenue-generating activities or for collaborations with commercial firms.

Commercial organizations seeking social innovation face problems similar to those of social organizations. Many organizations have reacted to the increasing scrutiny of their economic performance by developing cultures that value and reward market performance over the firms' other objectives, such as product quality and customer satisfaction. Such organizations may be very resistant to the adoption of "soft" social objectives as they are perceived to detract from commercial gain. To take advantage of opportunities for social innovation, these organizations need to move social objectives out of the realm of public relations, and accept them as increasingly important components of the organization's value proposition for customers and other stakeholders. Social values can become part of the organizational culture and identity in the same manner as commercial values, and should be tied into corporate socialization activities and values statements as an initial step toward incorporating them into policies and processes throughout the business.

Commercial organizations also encourage cultural perspectives that incorporate social innovation into the innovation portfolio. Encouragement of risk-taking and of allocating attention and resources to innovative activities promotes social innovation, just as it does commercial innovation when it is recognized that social outcomes are valued by the firm. Organizational mechanisms that support and reward innovation should be adapted to ensure that they accommodate innovation relating to the social values and objectives of the firm.

Strategy

Over time, every business model and technological characteristic in an organization loses its effectiveness in an ever-changing social and economic landscape. Therefore, social organizations must develop innovation strategies to ensure that they can maintain their effectiveness in accomplishing their mission and providing social value. The specific innovation strategy adopted by the organization must fit with the overall organizational strategy and values.

Innovation strategy depends on the organization's position within the sector it serves and the structure of its funding sources, along with its evaluation of the dynamism of the environment within which it operates. Like commercial firms, social organizations can choose to be "first movers" in generating and adopting innovations, or can follow behind the initial wave, adopting innovations that have proven benefits in the marketplace. The risk appetite and capabilities of the firm should be considered in developing the strategy.

To make the most of its innovative efforts, each firm must seek an appropriate balance between activities designed to promote idea generation and those designed to translate these ideas into action. Social organizations are often formed as means through which to address unmet social needs in innovative ways. Because the ongoing nature of many social goals ensures that market demand will always be available and that the threat from competing organizations is low, social organizations may not be as strategically focused as their commercial counterparts. In addition, sustained market demand, combined with a lack of resources that can be devoted to innovation, may encourage firms to prioritize increasing capacity over pursuit of new business models and methods. Adoption of a clear strategy toward innovation can help ensure that these organizations take advantage of technological and environmental opportunities that can contribute to the achievement of goals.

Social innovation strategies for commercial organizations are more complex than those in organizations with predominantly social missions. These strategies must be consistent with both overall corporate strategy and with the corporate strategy for commercial innovation. Due to competitive pressures, the need for continual improvement and innovation are well recognized in commercial organizations, yet many organizations lack a clear strategy to focus innovative efforts in a manner that best supports organizational goals. Because of the tendency to view socially oriented activities as competitors to activities that promote commercial outcomes, any social strategy must be carefully conceived and executed. The social innovation strategy must be built upon the core competencies of the organization, and the strategy must be carefully planned, such that existing competencies are strengthened or complemented by social innovations. Ultimately the firm may develop competencies in the social areas that support competitive advantage directly or enhance commercially oriented competencies.

Structure

Social organizations seeking innovation need to develop structures that allow innovation to flourish. Many structures are available, and innovation strategy should drive development of specific structures. In general, innovation is promoted by flattening and decentralizing organizational hierarchies. Structures that allow for coordination across functional units also promote exchange of knowledge and ideas within the company and enhance the environment for innovation within the firm. Dedicated innovation units that span other functional boundaries provide a means for direct support of innovation. Innovation is also supported by more fluid structures, such as temporary project teams, committees, and other collaborative efforts that can bring together managers from disparate units within the organization or connect managers with external stakeholders.

With innovation structures in place, social organizations can support innovations by developing internal markets for innovative endeavors. Organizations need organizational units that can identify high-potential innovations and provide the resources and support necessary to pursue them. These units can also monitor the effectiveness of innovative efforts and use the knowledge gained to refine innovation strategy and resource allocation decisions.

Commercial organizations that already have innovation structures in place can modify them or create additional structures to support social innovation. Firms can maintain existing structures and then modify them by incorporating social interests into innovation policies, resource allocation models, or evaluation processes. Alternatively, they can develop standalone structures dedicated to social innovation. To optimize social innovation and most effectively leverage it to promote commercial ends, these units must have strong and meaningful interaction with functional units to enhance a culture of innovation that spreads throughout the organization. The organizational level, visibility, and resources allocated to social innovation units serve to signify the relative importance of social pursuits and the hierarchy of organizational values.

In addition to effective internal structures, social innovation is heavily dependent on the existence of structures that promote relationships between the organization and its external constituencies. The social value generated by an organization is far more difficult to define and measure than its economic value. In general, a firm's stakeholders play an important role in defining and evaluating the social impacts of the organization. It is therefore imperative that a firm seeking social value develop structures for stakeholder consultation or engagement.

At a minimum, firms must look to stakeholders to determine the effectiveness of their social innovations. Stakeholders can be rich sources of knowledge and a broad range of other resources that can result in greater quantity and quality of social innovation. Organizational structures that span firm boundaries by establishing mechanisms for engaging stakeholders in operating and select strategic decisions can lead to ongoing generation of new ideas and opportunities. External stakeholders can play a role in evaluating the value-generating potential of social innovations for the firm, themselves, and other constituents potentially affected by the innovation. Additionally, they can help the firm anticipate and improve social outcomes associated with planned commercial initiatives. Ultimately, effective stakeholder engagement structures may result in the development of relational capital that allows for access to a network of resources that can be shared and accessed as needs arise.

Process

For social organizations, a primary component in social innovation is the business plan. Increasingly, large foundations, grant-providing organizations, and business collaborators require social organizations to adopt formal business planning processes. Organizations can benefit from a more rigorous approach to business planning that provides clarity in direction and detailed plans for mobilizing of human, material, and financial resources. Although funding and service levels plans may be well developed, social organizations may place less emphasis on planning for development of management skill, process alternatives, and new products and services—all strongly reliant on innovation.

In addition to standardized business planning, many social organizations can benefit from a more systematic approach to SWOT analysis—regular assessment of organizational strengths and weaknesses and their relationship to opportunities and threats in the environment. An important consideration in the analysis process is identification and development of the organization's core competency—the set of processes that contribute most to its competitive strength or advantage. To maximize the benefits of innovation, companies should allocate resources to enhance competence. Innovation in core processes and the related processes that most directly support them provides the greatest potential for continued strategic accomplishment and the creation of desirable resources. Innovation directed toward background processes such as accounting or legal compliance can divert resources away from strategic priorities.

To complement planning and competency development, organizations need processes to encourage ongoing improvement in business activities throughout the organization. Many social organizations fall far behind existing best practices of commercial firms in standard functional areas such as marketing or logistics. Social organizations need standardized ways of identifying best practices and incorporating them into operations.

For commercial organizations seeking to expand the role of social outcomes in the value proposition, business planning and competitive analysis should incorporate social objectives. Organizations need to develop an understanding of how competitors, customers, and suppliers are addressing societal and environmental issues and develop an understanding of the firm's relative strengths and weaknesses in the social realm. In addition, they need to develop processes to identify existing opportunities and threats and to anticipate and plan for future opportunities and threats. Rapid changes and increasing pressures relating to firm social performance ensure that firms with stagnant performance in this area will fall behind on this dimension, so firms need processes to ensure that innovation can accommodate social changes.

There is rapid growth in the social activity of commercial firms and in the resources available from supporting organizations. Among them are certifications for achieving external standards in environmental or social performance, standardized procedures for measuring and reporting social performance, and guidelines for effective stakeholder engagement. Organizations can benefit by adopting processes that allow them to capitalize on available external resources and using them as a foundation for driving and enhancing social innovation.

Firm approaches to basic process innovation should also be reconstructed so that both social and commercial capabilities can be engendered and enhanced. Commercial enterprises are often unaware of processes used by competitors and other peer firms to manage their own social and environmental outputs and outcomes. Organizations should engage in regular efforts to identify social best practices and explore how they can be implemented to enhance core competencies and further accomplishment of firm goals. Boundary-spanning relationships with stakeholders can be an important source of knowledge generation in support of innovation in this area.

Performance Measurement and Reward

The management tool that is perhaps used least effectively in the management of innovation is the performance measurement and reward system. For social organizations, performance measurement often focuses most heavily on the final service outcomes resulting from the organization's activity and funding inputs. Although this provides an effective means of evaluating past performance, social outcome measures provide insufficient information to manage and improve business performance.

Many commercial organizations fail to track and evaluate the effectiveness of the ideas generated, projects funded, or changes made as a result of their innovation efforts. Such measures can be useful in directing organizational attention and resources toward innovative endeavors. In addition, such measures can be used to identify successful and unsuccessful innovations, as a means for learning about the variables associated with success, and as a mechanism for communicating values and performance outcomes throughout the organization. The knowledge gained through performance measurement can be used to refine and improve innovation processes and their outcomes in the future.

Social organizations can also benefit from more rigorous attention to traditional financial reports. These reports are easily interpreted and can provide indications of both the efficiency with which the organization produces its product and the financial stability of the firm. This information can be used as a benchmark for internal or cross-firm comparisons and as a source of information about strengths and risks. In addition to traditional business measures, social organizations should directly measure and track their innovation efforts.

Reward systems can promote the firm's social innovation objectives. In social organizations, rewards can be linked to key social outcomes that represent accomplishment of the firm's mission. In addition, rewards can be used to encourage more directly the early actions that lead to social outcomes in the long run. Such actions might include idea generation, development of funding proposals, or successful proposal execution. Rewards can be used to motivate employees to take action that best supports execution of firm strategy and achievement of organizational goals. In cases where standard business practices have been overlooked, rewards can be used to redirect attention toward managing the commercial activities of the organization.

Commercial organizations are generally well equipped to measure and manage financial performance, but they often lack mechanisms to measure both innovations and social outputs. As with social organizations, measuring innovation activities and consequences can provide important information about each stage of the innovation process, and can be used as a basis for improving these processes and their outcomes. Organizations seeking social innovation should ensure that the innovation portfolio incorporates social objectives and is consistent with the company's strategy relating to social innovation. They can encourage social innovation by modifying standard measures of innovation effectiveness to accommodate social innovation. Ultimately, these measures can be used as means through which to monitor the contribution of social innovation to commercial performance objectives.

Unlike social organizations, commercial enterprises often have little understanding of the social outcomes associated with their operations. Often, they lack procedures for systematic monitoring—even of outcomes that are formally incorporated into the organization's values and objectives. For such firms, measures of social performance should be developed and incorporated into the performance measurement system to monitor and promote continued progress toward objectives. Measures should address social interests most critical to the firm as well as those of the organization's key stakeholders. In addition, areas that represent potential risks for the enterprise should be monitored. As with economic performance, the performance measurement system should incorporate leading and intermediate indicators of social performance along with measures of the outcomes and effectiveness of socially oriented action. In commercial organizations, social innovation can be encouraged by modifying standard measures of innovation effectiveness, such as ideas generated or changes implemented, to accommodate social innovations along with commercial ones.

Finally, measures of both social innovation and social performance should be incorporated into the organization's evaluation and reward systems. The relative emphasis placed on social performance in the reward system will influence attention directed toward social innovation. As the relationship between social and commercial performance develops and is more widely understood by organizational constituents, commercial performance measures can also be used to stimulate social innovation. Regardless of the outcome measures used, incentives and recognition directed toward innovation activity are also important, since the linkages between activities such as idea generation and commercial performance are difficult to trace, and the innovation cycle may be lengthy.

CONCLUDING REMARKS

To survive and thrive in today's dynamic environment, organizations of all forms must innovate—consistently and systematically. Continual innovation is necessary for protecting current sources of competitive advantage as well as for developing business models and technologies to maintain effectiveness in light of developments in the future. Commercial innovation alone is not enough. Organizations must embrace both commercial and social values to satisfy stakeholders and take advantage of emerging opportunities.

Social organizations face increasing demands from their supporters to demonstrate their efficiency and effectiveness in pursuing social outcomes. And they are beginning to recognize that methods used to generate innovation and advantage for commercial firms can be exploited to produce social benefits. Commercial organizations are likewise realizing that social innovation is key to producing outcomes that can be appreciated by a broad range of stakeholders and can simultaneously satisfy economic goals.

This chapter has provided a framework for exploring how social and commercial organizations can improve their social innovativeness by learning from each others' strengths. Social organizations can enhance their ability to achieve desired social outcomes by adopting the tools and disciplines of commercial innovation management. Commercial organizations can strengthen their vision and commitment to organizational values through the broadened perspective and creative insights that arise when an organization makes a positive impact on society. By adapting the guidelines presented in the chapter to fit their specific strengths and goals, organizations can improve their ability to innovate in a way that creates real economic, societal, and environmental value.

REFERENCES

Austin, James E. (2000) *The collaboration challenge: How nonprofits and businesses succeed through strategic alliances.* San Francisco: Jossey-Bass Publishers.

Bansal, P. and Roth, K. (2000) Why companies go green: A model of ecological responsiveness. *The Academy of Management Journal* 43: 717–736.

Barney, J. (1991) Firm resources and sustained competitive advantage. *Journal of Management* 17: 99–120.

Berman, S. L., Wicks, A. C., Kotha, S., and Jones, T. M. (1999) Does stakeholder orientation matter? The relationship between stakeholder management models and firm financial performance. *Academy of Management Journal* 42: 488–506.

Brinkerhoff, P. C. (1994) *Mission-based management: Leading your not-for-profit into the 21st century.* New York: Wiley and Sons.

Christman, P. (2000) Effects of best practices of environmental management on cost advantage: The role of complementary assets. *Academy of Management Journal* 43: 663–680.

Davila, T., Epstein, M., and Shelton, R. (2006) *Making innovation work.* Philadelphia: Wharton Business School Press.

Dees, J. G., Emerson, J., and Economy, P. (2001) *Enterprising nonprofits: A toolkit for social entrepreneurs.* New York: Wiley and Sons.

Emerson, J. and Twersky, F. (1996) *New social entrepreneurs: The success, challenge and lessons of non-profit enterprise creation.* San Francisco: The Roberts Foundation.

Mirvis, P. and Googins, B. (2004) The best of the good. *Harvard Business Review* (December): 1–2.

Nicholls, A., ed. (2006) *Social entrepreneurship: New paradigms of sustainable social change.* Oxford: Oxford University Press.

Paine, L. S. (2002) *Value shift: Why companies must merge social and financial imperatives to achieve superior performance.* New York: McGraw-Hill.

Porter, M. E. and Kramer, M. R. (1999) Philanthropy's new agenda: Creating value. *Harvard Business Review* (November/December): 121–131.

Schneitz, K. E. and Epstein, M. J. (2005) Exploring the financial value of a reputation for corporate social responsibility during a crisis. *Corporate Reputation Review* 7: 327–345.

Seelos, C. and Mair, J. (2005) Social entrepreneurship: Creating new business models to serve the poor. *Business Horizons* 48: 241–246.

Sharma, S. (2000) Managerial interpretations and organizational context as predictors of corporate choice of environmental strategy. *Academy of Management Journal* 43: 681–697.

Shore, B. (1995) *Revolution of the heart.* New York: Riverhead Books.

Steyaert, C. and Katz, J. (2004) Reclaiming the space of entrepreneurship in society: Geographical, discursive, and social dimensions. *Entrepreneurship & Regional Development* 16: 179–196.

Teece, D., Pisano, G., and Shuen, A. (1997) Dynamic capabilities and strategic management. *Strategic Management Journal* 18: 509–533.

Timmons, J. A. (1994) *New venture creation: Entrepreneurship for the 21st century.* New York: McGraw-Hill.

Waddock, S. A. and Graves, S. B. (1997) The corporate social performance—financial performance link. *Strategic Management Journal* 18: 303–319.

Wernerfelt, B. (1984) A resource-based view of the firm. *Strategic Management Journal* 5: 171–190.

8

Introducing Radically New Products and Services

ROBERT SHELTON

M any people talk about the need to introduce radically new products or services to create significant growth, but few really know how to do it successfully.

Most innovations that companies launch are incremental or breakthrough,[1] and very few are truly radical. Typically, in a given year, only about 5 percent of the innovation initiatives that are launched are radical, and of those very few move forward into subsequent stages of development. As a result, relatively few people have much experience developing and introducing radical innovations, and there is little shared or common wisdom regarding what it takes to successfully conceive and introduce radically new innovations.

Often companies launch a radical innovation initiative without having thought through how to manage the special requirement of radical innovations. Subsequently, they get stuck somewhere in the process of creating or commercializing the radical innovation and incorrectly decide that radical innovation is impossible for them. At that point, they abandon the concept of radical innovation, declare it off limits for their innovation investment portfolio, and artificially confine themselves to only incremental and breakthrough innovations. This constrained portfolio limits their competitive potential to something far less than it could be and cuts them off from the massive growth opportunities that radical innovation can provide. Or, worse can happen. Sometimes a company launches a costly radical initiative without adequate management, effectively betting the company on the radical innovation's expected payoff. Inevitably, without the correct management systems, the pay-off from the radical innovation fails to materialize, and the company loses its ability to compete effectively. At that point, the company either is acquired or goes out of business.

These types of negative experiences have led to many myths and misconceptions about radical innovation, such as:

- Radical innovation depends entirely on luck—you cannot manage it.
- Only mad-dog companies would invest in radical innovation—it is just too risky.
- Radical innovation is unique—it is not anything like incremental or breakthrough innovations.

Actually, radical innovation is not dissimilar to incremental or breakthrough innovation. It is a managed process that—with the correct inputs, metrics, incentives, and organization—can be combined to yield powerful, positive returns on investment. However, radical innovation cannot be managed in exactly the same manner as incremental innovation. It has some special characteristics and requirements.

TAXONOMY OF RADICAL INNOVATION

Radical innovations require significant changes to both the technology and the business models. This simultaneous change in technology and business models creates significant new businesses that typically have high growth rates and above-average profitability. The companies that successfully create and launch radical innovations are positioned to define the competitive environment in the new growth businesses. This is what provides the incentive for companies to embark on radical innovation—leadership and dominance of new growth business areas.

The current development of new space vehicles for tourism is an example of radical innovation. Taking tourists into space will require new, improved technologies that are different from the technologies that have traditionally been used to launch military or government projects. Tourists are not going to put up with the inconveniences and discomforts that the early space travelers had grown accustomed to. Safety will still be paramount, but comfortable seats, smooth takeoffs and landings, and reasonable toilets are probably just a few of the new requirements that space tourists will require. Likewise, the extremely high costs associated with government space travel must be drastically reduced to make it a viable option for tourism. This situation is not unlike early commercial airplane travel, where the airplane equipment, cabin comforts, and services that the airline companies provided were much different from the bare-bones military aircraft that preceded them.[2]

In addition, the business model for space tourism is markedly different from the historic government and military models. Government and military business models were focused on conducting research, increasing security, or improving military systems for defense. The selection of what went into space and when was tightly controlled and often highly secret. Launches were fairly infrequent. In addition, cost structure was secondary—delivering the service was the highest priority. This is massively different from space tourism. The business model for space tourism is focused on providing a safe adventure, but cost is a major consideration. It must be affordable. Also, access to space travel must be open to all who can afford it, and booking reservations will have to be convenient. In addition, there will need to be multiple flights per month—possibly multiple flights per week. These are major changes from the early days of military and government space travel and constitute the creation of an entirely new business. This radical innovation and the new business it provides has attracted attention from people such as Richard Branson, CEO of the Virgin business empire, who expect to create big, new, highly profitable space tourism businesses.

But radical innovations are not always related to rocket science—they can be much more down to earth. The combination of the iPod and iTunes is essentially a radical innovation because it created a new way to listen to music (the significant technology change associated with the user interface of the already existing MP3 player) and a significant new way to buy and access the music (a new business model that allowed, for the first time, the legal downloading of music from the Internet).

And what could be a more radical and yet down-to-earth innovation than disposable diapers? Baby diapers were traditionally made of woven cotton cloth, a technology approach to sanitary baby care that had not changed for a thousand years. Then, in the 1970s, the Scandinavians found that you could create absorbent diapers from fluff pulp derived from wood. The absorbent fluff pulp did the same job as the cotton, and both babies and mothers seemed pleased with the performance. However, cloth diapers must be laundered in the home or by a diaper service. Many families opted to use a diaper service that picked up dirty diapers and provided clean diapers once a week. The new diapers made from fluff pulp could be bought in the grocery store or other consumer retail outlets where mothers and caregivers frequently shopped. In addition, they could be disposed of at home. The new technology, combined with the new business model, was a radical innovation that created an entirely new category of consumer goods.

HOW DO YOU CREATE RADICALLY NEW PRODUCTS?

Developing and launching radical innovations cannot be done with exactly the same processes or people that it takes for incremental innovations. The nature of radical innovations and the need to simultaneously change the technologies and the business models make the development process somewhat different than what is required for incremental innovation. If you think of sailing as a simile for innovation, then it is fair to say that radical innovations are to incremental innovations as sailing in the open sea is to hugging the shore.[3] Both types of sailing have risks and uncertainties, and both require the same basic equipment. However, the lack of a visible shoreline requires slightly different sailing tools and a greater ability to handle risk and uncertainty.

In a similar manner, managing radical innovation is similar to incremental and breakthrough innovation management, but it requires a different approach. Incremental innovations make modest changes to the existing technology and business models, but never venture far. The existing technologies and business models are always in plain sight and guide the development of the small changes. Breakthrough innovations make a significant change to either the business model or the technology but essentially maintain one or the other in its original state. The essentially unchanged element provides a guide that aids the development process—a visible shoreline to aid navigation of the innovative venture.

Radical innovations do not have that shoreline; they make significant, simultaneous changes to both the technology and the business model. Managing radical innovation requires leaving the guidance and comfort of the known, visible shoreline in search for something new.

THE 3 PS OF RADICAL INNOVATION

Radical innovation requires management of the 3 Ps:

- People
- Partnerships
- Processes

The 3 Ps are as valid for managing incremental and breakthrough innovation as they are for managing radical innovation. However, there are a few special twists to the 3 Ps for radical innovation.

People

The people associated with radical innovation are not the types of folks you would hire to make small incremental changes to a product. Developers of radical innovations are a special breed; they are explorers with an undeniable desire to see their dreams turned into reality. In addition, the innovators are so filled with inspiration and desire that they are insufferable—it is all they can think about, talk about, or focus on.

Consider Columbus, one of the best examples of this type of innovator. Columbus had a different view on the best way to reach India. His vision of sailing west to get to the Indian spices was met with skepticism and even mockery. This did not deter him, although it might have stifled a man of less conviction. By all accounts, Columbus made a pest of himself in the courts across Europe, trying to raise capital to finance his idea. Not only were his ideas counterintuitive, but his demeanor was difficult to stomach as well. His unwavering belief in his vision of "Go west to go east" and his unshakeable confidence in his ability to sail the distance in the uncharted open seas made him argumentative, stubborn, and difficult. He had a vision of where to sail, and he knew he had the ability to sail into the uncharted open sea and to make it happen. After many years, Spain's Ferdinand and Isabella gave him the funding he had been badgering them for. It may have been a good way to get rid of the pest, or else the most inspired piece of venture funding—we will never know. However, we know that radical innovations need people like Columbus to lead them—visionary, excellent at execution, and unbelievably stubborn.

How do you identify a radical innovator? Above all, the radical innovator has to want to change the world. Guy Kawasaki says that from all his experience with great innovators—and Kawasaki has been a leader within Apple Computers as well as a venture capitalist (VC) and start-up guru in Silicon Valley—the really effective innovators do not say "I want to make more money." Those money-oriented innovators are not the ones who can envision really major innovations or lead the team to success.[4] The leaders of radical innovation initiatives absolutely must see world-changing potential in what they are trying to do. Rudolf Diesel, inventor of the diesel engine, envisioned a device that would change society significantly by providing clean, cheap energy. His vision was not just to build a better internal combustion engine, it was to make a difference in the way the world used energy and created wealth. He wanted a world with far greater equality and less social strife, and he saw his engine as a vital force in making that happen.[5] To him the innovation was a means to a far greater end. Radical innovators always have that sort of grand belief. So when you are looking for members for your radical innovation team, ask candidates to tell you their aspirations for the initiative. If they do not respond "To change the world," don't select them.

Partnerships

It is important to remember that innovation is a team sport. It is not dependent on just one person—not even the visionary leader. Innovation requires a large range of talents and levels of effort that are more than any one person can deliver. It is more than the core innovation team can deliver. Innovation, especially radical innovation, depends heavily on a network of partnerships that provide a collaborative business ecosystem to create and commercialize the innovation.

Some companies are good at changing both technology and business models. GE, FedEx, and Apple have demonstrated the ability to manage change in both areas. However, that is not always the case. Many companies have excellent technology innovation capabilities and can push the boundaries of technology to new frontiers. However, they are less adept at significantly changing business models to create new value propositions and delivery approaches. Other companies face the opposite challenge; they are good at business model change but not as adept at technology change. Dell has historically been a company that has worked business model changes well but has not shown the ability to effect commensurate changes in the technology arena.

Partnerships are an excellent way for radical innovators to have a balanced capability and approach. Companies that lack the necessary strength in either business model change or technology change must establish collaborative partnerships to fill in their gaps. This is critical in radical innovation when success depend on significant, simultaneous change to both the business model and the technology. It may even be appropriate to use nonconventional partnering approaches such as open-source collaboration. Open-source software development projects—Internet-based communities of software developers who voluntarily collaborate in order to develop software—have become important economic and cultural phenomena and exemplify the extent to which partnerships can be used in innovation.[6] For radical innovations, it may be entirely appropriate to use such collaboration techniques.

Radical innovations are inherently risky, and managing risk is a major responsibility for the leadership team. That includes demonstrating to funders and stakeholders that the potential benefits of the project far outweigh the costs and risks. To support the radical visionary in acquiring funding, it is best to build a world-class advisory team to help define the ideas of the visionary and then sell them to others. There is strength in numbers. People react to ideas differently when there are multiple, credible sources. The radical innovation team should try to assemble a team of world-renowned supporters because their credibility will be vital. However, it is most important to have people who believe in the idea and the innovator. "I have built a team who have believed in me and seen me pull off a number of crazy ideas," says Peter Diamandis. Mr. Diamandis, the entrepreneur who put together a $10 million X Prize for the first successful private sector space flight in 2004, sold the concept of the X Prize by gathering twenty astronauts as supporters. And when he pitched his idea for an International Space University, a company that offers zero gravity space flight, he included a former secretary of the Air Force, a key leader from NASA, and a two-time shuttle astronaut. "They spoke our ideas with such credibility." [7]

Processes

Most of the processes required for radical innovation are generally the same as for incremental and breakthrough innovations. For example, all types of innovation require a balance between creativity—creating the great new ideas—and commercialization—bringing the ideas to commercial reality. However, it takes a special combination of people, partnerships, and processes to create and commercialize radical innovations.

For one thing, radical innovation requires different management processes. For example, measuring progress for incremental innovation is easy; it merely requires checking whether project milestones are being met. In contrast, measuring progress for radical innovation is based on more subjective evaluations of whether the experiments and prototyping is producing valuable learning. In radical innovation, it is more important to assess how much learning is being accomplished and how well it is being used than to measure progress against an assumed outcome—an outcome that changes constantly based on new learning. So instead of measuring the progress against the assumed outcome, progress is measured by the amount of learning that is generated to define what the technical performance and business models will have to be to achieve success.

Highly formalized and rigid management systems are a deterrent to radical innovation. Radical innovators are goal driven, not process driven. For that reason, a venture capital model is used for radical innovation. The venture team members bring together their experience and instinct and use a flexible system to guide the project. At the early stages of development, the team favors creativity and learning. In the later stages, the team focuses more on rigor of analysis and financial measures to guide decisions.

Managing radical innovation is driven primarily by achieving high-quality team dynamics and sustained levels of robust collaboration. Likewise, the metrics and incentives for radical innovation are focused on the inputs to the process rather than the outputs. Often the results of radical innovation are not known for a long time after the project is completed. Therefore, it is not possible to measure and reward performance based on results. Instead, metrics should focus on the inputs to the process (such as the quality of the people and the effectiveness of the partnerships). Continual support from management and recognition are two of the most powerful rewards for the radical innovation teams. Trying to buy radical innovation is doomed to fail. Remember that radical innovators do not work hard for money; they are out to change the world. Their vision—not the quest for financial compensation—fuels their drive to succeed.

Incremental innovation uses knowledge management of technologies, markets, and business models to guide development. Mining rich databases and using the cumulative knowledge of the organization is a key to success. That is not the case for radical innovation. Innovation processes for radical innovation must engage in aggressive ignorance management to succeed. At the onset of a radical innovation project, there is an inherently large amount that is not known; in fact, because it is a radical innovation, more is unknown than known about the new markets, the performance of the technologies, and the adequacy of the business model. In this type of situation, it is crucial to experiment and learn what you need to know. When Salesforce.com launched its concept of managing sales forces and customer management through the Internet, there was little knowledge to rely on. Little was known about how medium- and large-sized companies would use the system and which features would be most valuable to them. Even more importantly, it was not known if the business model using the Internet would be sufficiently attractive to lure customers away from the traditional approach of buying the software and installing it on the company server.

Ignorance management is the process of managing what you do not know, identifying the critical pieces of information that needs to be found, and finding fast and effective ways of getting that information. Often companies will use trial and error and prototypes as part of their ignorance management approach to probe and learn. Salesforce.com used trial-and-error and prototyping approaches to develop information about the markets and technologies. The company gathered data on how customers used the service and the faults with the business model. They used this to make quick fixes, and later in the development cycle they mined that data to help them identify other areas for improvement. Rapid prototyping and learning should be core competencies for every radical innovation team.

CONCLUSION

Radical innovation has special requirements. The most common mistake is using the same management processes for radical innovations as are used for incremental innovation. Radical innovation has different management requirements for people, partnerships, and processes. Successfully managing these elements will create significant competitive advantage and growth.

NOTES

1. For a complete description of the types of innovation—incremental, semiradical or breakthrough, and radical—see *Making innovation work* by Davila, Epstein, and Shelton, Wharton Business School Publishing, 2005.

2. Schwartz, John. 2003. Into space, without NASA. *New York Times*, August 26. page D1.

3. Robert Shelton, "Making innovation work for sustainable business: How to manage it, measure it, and profit from it," Commonwealth Club, San Francisco, December 13, 2005.

4. www.alwayson.com.

5. The fact that the diesel engine did not bring about the level of social change that Diesel wanted was a major disappointment and ultimately led to his death.

6. *Making innovation work*, page 103.

7. *Still crazy after all those light years*, Mark Turner, *Financial Times*, October 5, 2005.

9

Turning Creativity into Value Creation: The Growth Path of Start-up Firms

BERNHARD R. KATZY and FLORIAN STREHLE

To be successful, innovative enterprises have to turn creativity into value creation. In this chapter, we are concerned with the growth path of start-up firms, the process of a venture that sets off mainly with a creative idea and turns it into a profitable business over time. Consequently, we take growth as a dynamic process and seek to understand which factors determine its success. We choose an inside-out perspective to identify critical skills that entrepreneurial ventures need to succeed in driving the process of organizational growth.

The study sets out from the assumption that the capability to manage an organization along its growth path is a resource in its own right. This capability complements other resources like innovative technology, a creative product or service, or a superior production process, which are equally necessary for science- and technology-based new ventures. The first contribution of the study is to identify and name concrete entrepreneurial resources and empirically observe their impact over time on the growth process.

If entrepreneurial management is a resource, it should develop and accumulate over time, just as technology is built through ongoing research and development and production processes evolve along the learning curve. The second contribution of this chapter, therefore, is a model of capability building in new ventures. The study contributes empirical evidence on the evolution of the firm while it grows. It is especially possible to operationalize evolutionary conceptual foundations. Besides its analytical value, the list of identified entrepreneurial capabilities and their descriptions contribute to practicing entrepreneurial managers' guidance and "best practice" indications.

DYNAMIC CAPABILITIES AS THE FIRM'S GROWTH ENGINE

The resource-based view as a strategic management framework originated in the work of British economist Edith Penrose and her seminal publication on "the theory of the growth of the firm" in 1959, in which she posed the thesis that a firm grows to the limit of its managerial abilities. Broader perception of her work, however, did not start until almost forty years later. In 1984, Birger Wernerfelt referred to Penrose's ideas when he argued that the evaluation of firms according to their resource endowments could lead to insights that differ significantly from traditional approaches in strategic management. And it took his paper another ten years to receive a best paper award. In this perspective, organizations are considered as bundles of resources (Peteraf 1993; Eisenhardt and Schoonhoven 1996). In general, resources can be defined as stocks of available factors like machines, capital, or patents that the organization owns or controls (Amit and Schoemaker 1993), as well as learned patterns of organizational routines. Because this way of strategic reasoning starts with the internal characteristics of the organization of the firm, the concept is often referred to as the "inside-out" counterpart to the mainstream industrial organization "outside-in" perspective (Porter 1980; Porter 1985), where the firm's structure follows its strategy (Chandler 1962) in its external environment.

In contrast to large established firms that own a wide range of valuable resources, entrepreneurial firms, by nature of being new, do not. This can be an advantage in situations of radical innovation, where the former valuable resources become rigidities for established enterprises (Leonard-Barton 1992) or competence traps (Levitt and March 1988). But young firms are challenged by resource scarcity in the beginning and have to establish a proper resource base in order to implement the new business idea and grow (Stinchcombe 1965; Romanelli 1989; Brush, Greene, and Hart 2001; Stuart and Sorenson 2003; Ravasi and Turati 2005). Stuart and Sorensen (2003) identified three different types of resources that are most critical for launching a science and technology-based venture. First, start-ups need a new idea or foundational technology. These, however, are a dime a dozen, and sometimes different entrepreneurs even come up with almost the same plan at the same time (Bygrave and Zacharakis 2003). Rather than gauging success by the idea itself, the success of a new venture is determined by the entrepreneur's ability to recognize promising opportunities. We therefore treat this ability as a valuable resource in its own right (Alvarez and Busenitz 2001). A second type of resource is funding (Schoonhoven, Eisenhardt, and Lyman 1990; Klofsten, Jonsson, and Simón 1999; Miller and Garnsey 2000; Lounsbury and Glynn 2001; Stuart and Sorenson, 2003). Inadequate financial resources are a major cause of young business failure (van Auken and Carter 1989). Entrepreneurs need fundraising ingenuity

(Penrose 1959) if they depend on external financing (Evans and Jovanovic 1989) like venture capital (Gompers and Lerner 1999). Employees with highly specialized human capital are a third critical resource for new ventures (Kamm et al. 1990; Lounsbury and Glynn 2001; Stuart and Sorenson 2003). Various scholars (e.g., Prahalad and Hamel 1990; Nelson 1991; Nonaka 1991; Henderson and Cockburn 1994; Nonaka 1994; Kogut and Zander 1996; Nahapiet and Ghoshal 1998) claim that knowledge and the ability to create and apply this knowledge are the most important sources of competitive advantage in both highly dynamic environments and environments with disruptive change (Grant 1996)—which are the most suitable environments for entrepreneurial ventures.

In summary, a crucial task of the entrepreneurial top management team is to build the resource base for the new firm. If it is true that especially in technology start-ups, founders often lack the necessary business acumen (Roberts 1991; Brush, Greene and Hart 2001), this top management capability will be a factor to distinguish successful and less successful new ventures.

The general top management capability can be detailed into a number of specific tasks: superior founding teams have the capability to attract potential investors. Venture capitalists usually apply the quality of the firm's founding team as an important evaluation criterion for their investment decision (Tyebjee and Bruno 1984; MacMillan, Siegel, and Subbanarasimha 1985) and rather compromise on the presented business idea if they have the chance to invest in superior teams (Muzyka, Birley, and Leleux 1996; Wright and Robbie 1998). And strong management teams do provide an entrepreneurial venture with access to resources such as specialized labor or social capital resources (Ucbasaran et al. 2003). In summary, rather than a lonely entrepreneur as the "jack of all trades" (Galbraith 1982) who covers many functions but often masters only a few of them (Schoonhoven, Eisenhardt, and Lyman 1990), organizational entrepreneurial capabilities of functional experts in coordinated teams should have a positive influence on the growth of a new firm.

THE TIME DIMENSION OF GROWTH PROCESSES

Explicit consideration of the time dimension is the second intertwined conceptual contribution associated with the resource-based view of the firm. The pace of change in turbulent environments (Bourgeois and Eisenhardt 1988; Eisenhardt 1989) makes it more and more difficult to sustain an advantageous position. Instead, advantageous positions have to be recreated constantly. And actually, the nature of entrepreneurial ambition is to drive the process of creative destruction (Schumpeter 1934) and change.

Barney (1991) offers a conceptual bridge to mainstream strategic literature when he distinguishes between time-independent competitive advantage that stems from a good fit of the firm with requirements of its industry on the one side and sustainable competitive advantage over time. The latter requires that competitors cannot quickly duplicate the benefits of the firm's strategy, for example, through buying the same machines or technology. Homogenous resources and resource mobility work against sustainability of competitive advantage (Barney 1991). In contrast, resources that can be considered immobile are those that cannot be traded (Peteraf 1993) but instead are created internally or, in the case of organizational routines, are learned over time. In this regard, the time it takes to build a valuable resource is a source of advantage and an inhibitor to strategic fit.

Scholars have developed a wide variety of approaches and theories to explain development and change in organizations. Especially life cycle, stage-of-growth, and evolutionary models are applied regularly to denote the growth processes of firms (Gruber, Harhoff, and Tausend 2003). All of these approaches imply that firm growth follows predictable patterns that occur at discrete periods of time (Smith, Mitchell, and Summer 1985) and that actions taken by management with respect to current problems drive the transition to the following stage (Dodge and Robbins 1992). Contingent upon the current life cycle stage, firms differ along various dimensions such as age, size, targets, structure, control, communication, leadership, key personnel, reward systems (Greiner 1972 1998), or changes in strategies, priorities, problems (Kazanjian 1988), politics (Gray and Ariss 1985), or formalization (Walsh and Dewar 1987). In short, at different times in the growth process, priorities of growth management vary.

Similar to life cycle models, stage-of-growth approaches use discrete stages to explain organizational development. But in contrast to the assumption of homogenous evolution in life cycle, models stage-of-growth models assume a discontinuous development of the firm, which is shaped by crises. Influential stage-of-growth models for entrepreneurial ventures comprise the approach by Churchill and Lewis (1983) and Galbraith's (1982) five-stage model, which explicitly addresses technology-based new ventures. In other words, in these models, the process is denoted as a sequence of crises situations that growth management has to cope with over time.

Evolutionary models combine the characteristics of life cycle models with those of stage-of-growth approaches. Evolutionary models consider organizational crises and the corresponding influences on the development of the firm (Gruber, Harhoff, and Tausend 2003). Important evolutionary models for organizational growth and change comprise Scott and Bruce (1987) as well as Greiner (1972, 1998). Although Greiner's work is one of the earliest models for organizational development (van de Ven and Poole 1995), it is considered a baseline in this field (Hanks et al. 1993). Growth, here, is a learning process over time, which is triggered by intensive experiences and to which entrepreneurial management reacts depending on their capabilities.

GROWTH AS A PROCESS OF REDUCING UNCERTAINTY

The reason that many new businesses fail can often be found in the high levels of uncertainty that they usually face (Bourgeois and Eisenhardt 1988;

Eisenhardt 1989). Uncertainty is "a lack of predictability, of structure, of information" (Rogers 1962, 6) and has to be clearly distinguished from risk. In case of risk, ex ante calculations can be performed that indicate the probability for an event to occur. This is not possible for uncertainty. Uncertainty is a key component of entrepreneurship. In fact, only uncertainty can explain extraordinary profits as well as failures of entrepreneurial firms (Knight 1921; Brouwer 2002). Sources of uncertainty are manifold in entrepreneurial ventures. Rapid change in the environments makes it impossible to predict market demand, technology development, and changes in competition and regulation (Bahrami and Evans 1989), and thus the necessary combination of resources that fit.

The ability to manage the growth process is an ability to reduce uncertainty and to structure and restructure the firm. It therefore is a dynamic capability (Eisenhardt and Martin 2000) that enables firms to continually acquire, upgrade, and develop resources in order to grow and maintain a competitive position in the market (Wernerfelt and Montgomery 1988) or adapt to emerging new settings with minimal effort and short time delay (Nelson and Winter 1982; Hayes and Pisano 1994). Eisenhardt and Martin (2000) define dynamic capabilities as a "firm's processes that use resources—specifically to integrate, reconfigure, gain, and release resources—to match and even create market change. Dynamic capabilities thus are the organizational and strategic routines by which firms achieve new resource configurations as markets emerge, collide, split, evolve, and die" (p. 1107).

Organizations that are more mature and less entrepreneurial should face lower levels of uncertainty. And because effective capabilities must match the degree of uncertainty and entrepreneurial dynamics (Deeds, DeCarolis, and Coombs 1999; Eisenhardt and Martin 2000), firms need either to continuously adapt their existing routines or replace them with new ones. Consequently, advances in growth are not only more of the same routines, but should become visible as changes of routines. The empirical part of this study thus focuses on which routines and capabilities are developed by growing new ventures.

For science- and technology-based new ventures, a variety of dynamic capabilities are important in all management functions: financial, strategic and human resource planning, financial and human resource evaluation (Schuler and MacMillan 1984), product development (Grant 1996; Deeds, DeCarolis, and Coombs 1999; Rangone 1999; Verona 1999; Rothaermel and Deeds 2005), sales and marketing (Verona 1999; Shepherd, Douglas, and Shanley 2000; Gruber 2004), as well as partnership management (Minshall 1999; Minshall 2003; Rothaermel and Deeds 2005). Planning capabilities are necessary for new ventures since planning reduces uncertainty (Armstrong 1982; Grinyer, Al-Bazzaz, and Yasai-Ardekani 1986; Smeltzer, Fann, and Nikolaisen 1988; Shrader, Mulford, and Blackburn 1989). Figure 9.1 summarizes the different dynamic capabilities.

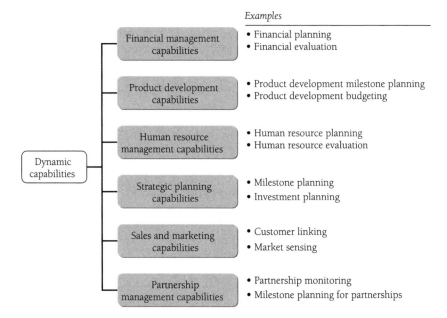

FIGURE 9.1. Overview of Dynamic Capabilities in Technology-Based New Ventures

PATH DEPENDENCY OF ENTREPRENEURIAL FIRMS

Growth is presented here as multiple intertwined processes of developing resources and routines, which sum up the specific, difficult-to-imitate configuration of the firm that is the source of sustainable competitive advantage. We have identified some general capabilities in the previous section that apply to all entrepreneurial ventures. The difference between firms stems from the individual shape that the web of organizational routines takes in the course of the growth path. This is in line with many scholars (e.g., Nelson and Winter 1982; Teece, Pisano, and Shuen 1997; Zollo and Winter 1999) who argue that capabilities stem from path-dependent processes (Eisenhardt and Martin 2000).

The notion of path dependency is important since it explicitly incorporates the influence of a company's history on the evolution of capabilities. A more accurate denotation was given by Eisenhardt and Martin (2000), who argue that learning mechanisms drive the evolution of dynamic capabilities in organizations. Capabilities can either be imitated from other firms or developed from scratch. In both cases, organizational learning is necessary since the new venture has never engaged in this particular activity before (Helfat and Peteraf 2003).

Managing the growth of entrepreneurial ventures can be seen as a sustained investment in learning efforts. Internal to the organization, high growth implies an increase of organizational size and complexity in a short period of time, which pushes firms to develop corresponding capabilities (Sexton et al. 1997). External to the organization, high-velocity entrepreneurial environments push the firm to renew capabilities in reaction to market dynamics.

For the purpose of our study, progress in the learning process is an indicator for sustained growth of the firm. This process has been studied in its own right and was detailed in four phases: knowledge acquisition, information distribution, information interpretation, and knowledge codification (Fiol and Lyles 1985; Levitt and March 1988; Huber 1991; Day 1994; Sinkula 1994; Bierly and Chakrabarti 1996; Kloot 1997). Because we are interested in the effect of learning on the growth process—rather than the learning itself—we go directly for the last phase of knowledge codification. Organizational routines and daily operations are important repositories of firm knowledge, as are documents, manuals, specifications, patents, or databases. In fact, a large amount of a firm's knowledge about how to perform tasks is stored in its operating routines (Nelson and Winter 1982; Huber 1991). These knowledge assets are not tied to certain individuals and consequently can be retained if key people leave the firm (Kazanjian 1984; Kogut and Zander 1992).

As we are mainly interested in management routines, management control systems (MCS) are an especially interesting class of codified knowledge. MCS are "formal, information-based routines and procedures used by managers to maintain or alter patterns in organizational activities" (Simons 1994, p. 170). They are necessary for adaptive learning in organizations since they are important knowledge repositories (Davila 2005). They measure the gap between target and actual outcomes within operating routines and thus influence the efficiency of organizational processes. Organizations apply management control systems to gather and use information, which supports planning, and to control decisions throughout the firm. MCS are often supported by management information systems (Lorange and Scott-Morton 1974), are subject to documentation, and therefore can easily be observed, as in Table 9.1. Management control systems are also interesting because they are means to reduce uncertainty (Tushman and Nadler 1978). Since growth is associated with the reduction of uncertainties (Dissel 2003), sustained growth can be expected to be associated with the continued creation of MCS.

Learning, of course, is not identical to firm performance. Figure 9.2 shows the links between learning, dynamic capabilities, and routines in the new venture. While dynamic capabilities evolve from learning efforts, they drive the development of primary processes and operating routines, which create economic rents.

The task of the entrepreneurial top management team is to balance the three levels. Entrepreneurial firms that neglect primary processes are not likely to perform with regard to economic rents. Firms that hold on to traditional practices for too long are prone to severe crises or failure. Instead, organizations must find new management patterns that serve as a basis for the next period of evolutionary growth (Greiner 1972, 1998).

TABLE 9.1. Classification of Management Control Systems

	Management Control System
Financial planning capability	Cash-flow projections Sales projections Operating budget
Strategic planning capability	Investment budget Definition of strategic (non-financial) milestones Product portfolio plan Customer development plan (plan to develop the market) Headcount/human capital development plan
Human resource planning capability	Core values Mission statement Organizational chart Codes of conduct Written job descriptions Orientation program for new employees Company-wide newsletter
Financial evaluation capability	Capital investment approval procedures Operating expenses approval procedures Routine analysis of financial performance against target Customer acquisition costs Customer profitability analysis Product profitability analysis
Human resource evaluation capability	Written performance objectives Written performance evaluation reports Linking compensation to performance Individual incentive programs
Product development capabilities	Project milestones Product concept testing process Reports comparing actual progress to plan Project selection process Product portfolio roadmap Budget for development projects Project team composition guidelines Product development monitoring system Product quality monitoring system
Sales and marketing management capabilities	Sales targets for salespeople Market research projects

	Sales force compensation system
	Sales force hiring and firing policies
	Reports on open sales
	Customer satisfaction feedback
	Sales process manual
	Sales force training program
	Marketing collaboration policies
	Customer relationship management (CRM) system
Partnership management capabilities	Partnership development plan
	Partnership development reports
	Policy for partnerships
	Partnership milestones
	Partner monitoring system

(Based on Davila and Foster 2005)

METHODOLOGY

The study empirically tests the growth process with technology-based high-growth new ventures in the Munich region. We identified eighty-eight venture capital-backed companies because more information is available from both the firm and the investor, and VC-funded firms usually show higher growth rates. Forty-four firms replied positively and participated in the study. Out of the forty-four, the smallest firm had ten employees, while the largest one accounted for 250 at peak time. The average peak number was fifty-three employees. The youngest firm was three years old, the oldest twelve years old, with an average of 5.7 years. Peak revenues were between €0[1] and €26 million, with a mean of €4.7 million. VC investment ranged from €0 to €61.4 million, with a mean of €15.5 million. All participating companies were asked to complete a questionnaire detailing which MCS had been introduced in which year since their founding. Further, we collected general company information, such as financial figures and critical events, especially as indicators for crises situations. In addition, we conducted one to three interviews per firm and considered publicly available information as well as a commercial database on VC-backed firms to triangulate the information obtained.

Because capabilities are complex configurations of processes, routines, and tasks, each of which can make use of a different management control system, we introduced intensity measures for each capability as the percentage of observed MCS in a particular group. For example, the financial planning

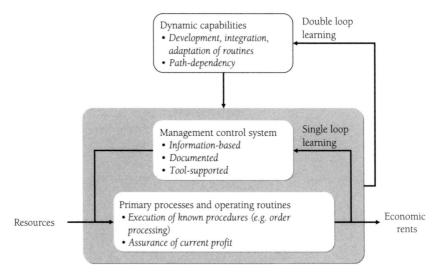

FIGURE 9.2. The Links between Learning, Dynamic Capabilities, and Routines

capability includes budgeting as well as sales and future cash flow projections. We approached learning and evolution of any capability as an increase in MCS usage over time. Some MCS also contribute to different capabilities. A MCS is only considered as existing if the respective routine or structure is either documented in written form or repeatedly and purposefully executed and is organizational rather than individual in that it is executed by more than one person. Figure 9.3 shows the increase of MCS adoption over time.

More interesting in Figure 9.3 is that not all capabilities evolve at the same time and with the same intensity. Two reasons could account for this. On the one hand, the development of certain skills might require more time than the evolution of others. On the other hand, it could be possible that some capabilities are not required in the very early days, while others are necessary to drive the growth of the firm from the start.

To further elaborate on this question, we compared the firm growth rates with the existence of a certain capability as independent variable. We categorized the sample companies into two groups for each capability: the early adopters and the late adopters. The same MCS intensity method was used to cluster on the basis of capabilities, not individual MCS. The longitudinal nature of the process was measured by grouping the firms in several years since foundation and relating the capability to firm growth rates, which are measured in number of employees. Mean growth for a certain year was calculated as the number of employees at the end of each year divided by the number of years. For the first three years, we have forty-four observations. In year four, only thirty-eight firms can be considered.

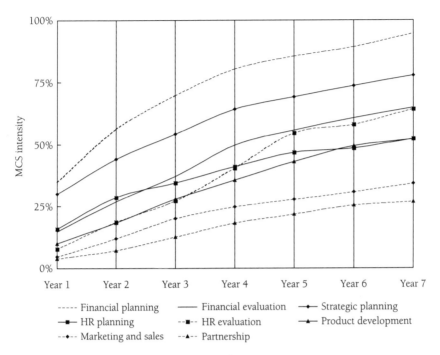

FIGURE 9.3. Increase of MCS Intensity over Time

RESULTS

Financial Planning Capability

This type of MCS includes the development of operating budgets, cash flow projections, and sales projections. We clustered the sample companies in the first four years according to their MCS intensity in this particular field. Figure 9.4 compares mean employee growth rates in the first five years. The early and late adopter firm groups are distinguished based on whether they adopted MCS in the second year.

Firms with high system intensity show more than double the growth rates of new ventures, which have a low MCS intensity in terms of financial planning. The results are not very surprising. Financial resources are highly important, but scarce in the early days. Thus, the allocation of funds has to be planned carefully. Growth differences between the two firm groups get bigger starting from year three, which gives an indication of about two years' time lag between building financial planning capabilities and observable impact on the growth of the firm.

Financial Evaluation Capability

This capability comprises the analysis of product and customer profitability, customer acquisition cost analysis, or certain rules for the allocation of funds.

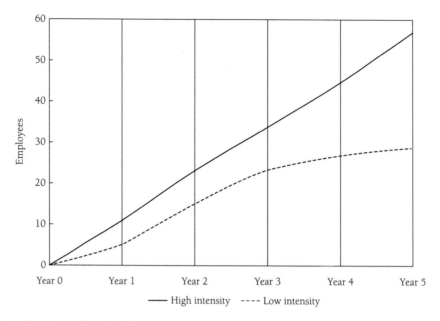

FIGURE 9.4. Financial Planning MCS Intensity in Year Two and Firm Growth

Again, we seek to identify an association between MCS intensity and firm performance, i.e., an increase in employees. Again, we grouped companies in the
first four years and denoted the corresponding growth rates of both groups.

Interestingly, for the first two years group one underperforms, i.e., firms
that heavily adopt financial evaluation systems in years one or two show
lower growth rates than firms that do not. A possible explanation is that in
the very early days, start-ups do not sell much or lack a large customer base.
Consequently, financial evaluation capabilities are not yet necessary.

The picture looks different for years three and four. The latter results are
denoted in Figure 9.5. Now group one performance is better. This is in line
with stage models of growth that predict different management priorities for
different phases. While a new venture initially engages in technology development, a firm that is two to three years old could have sold its first products
to early-adopting customers. Consequently, in this phase financial evaluation
skills are necessary to evaluate the corresponding success.

Strategic Planning Capability

Strategic planning comprises the development of an investment budget,
certain strategic milestones, or a product portfolio roadmap. Again, groups
were built for the first four years.

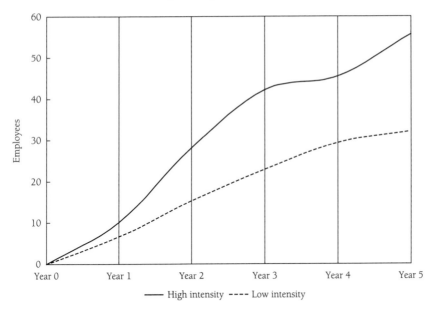

FIGURE 9.5. Financial Evaluation MCS Intensity in Year Four and Firm Growth

Clustering the companies in the first two years indicates a positive associa-tion between strategic planning MCS emergence and new venture perform-ance. The results for the two groups determined in year two are shown in Figure 9.6. Categorizing the firms in years three and four shows a different picture. Now both groups of companies indicate almost identical growth rates. Apparently, available strategic planning capabilities are especially critical in the early beginning. Interestingly, the impact of the early existence of these capabilities does not become visible in the first three years. Although clus-tered in year two, firms with high intensity in terms of strategic planning MCS first show substantially higher growth rates in year four. A reason for this outcome could be the character of uncertainty in start-up firms and its resolution through planning activities. Since technology-based new ventures face high levels of uncertainty in various fields, it takes substantial time and effort to reduce the uncertainty. Firms that do not initially invest in this effort might encounter much smaller growth rates than companies that reduced uncertainty from the very beginning.

Human Resource Planning Capability

This group comprises the development of a mission statement, core values, an organizational chart, or the offering of orientation programs for new employees. We group the sample firms in the first four years according to

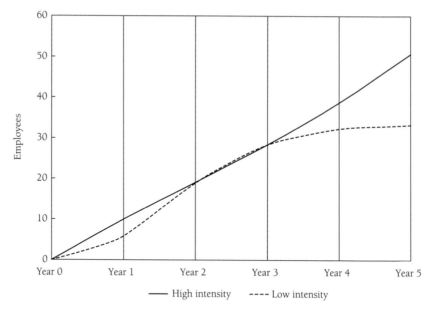

FIGURE 9.6. Strategic Planning MCS Intensity in Year Two and Firm Growth

their human resource planning MCS intensity. The employee increase of both groups defined in year three is indicated in Figure 9.7.

Comparing the growth rates of both groups in year one reveals comparable results except for the first year. The annual mean growth based on firm size in years two and three is even lower for high-intensity companies. However, the picture changes with time. While the grouping in year two already reveals a slightly better performance for high-intensity companies, the effect becomes stronger for years three and four. For the last two years of grouping, the difference in growth between adopters and nonadopters of MCS becomes substantial.

In the first year of existence, a new venture often comprises only the management team. These people usually do not need orientation programs, and written job descriptions will not add any benefits to the company since most organizational members assume the roles of jacks of all trades. Thus, the capabilities to implement and adapt these elements are not necessary. However, as the company becomes larger, the tasks become more specialized and focused, and organizational structure assumes a more important role. In our sample, the mean size in year three is twenty-eight employees. At this time, new ventures can no longer be managed in an informal and unstructured way. Now the executive team must be able to manage the employees in a different way and thus has to develop capabilities to establish structures and

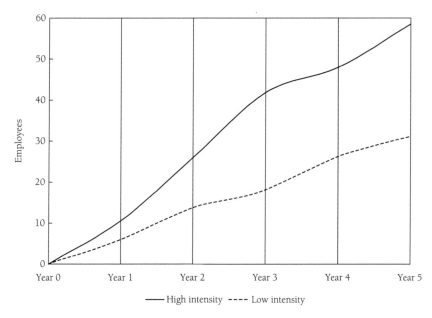

FIGURE 9.7. Human Resource Planning MCS Intensity in Year Three and Firm Growth

guiding principles manifesting in the introduction of organizational charts, job descriptions, or a company mission and vision.

Human Resource Evaluation Capability

The next group of human resource MCS covers the field of personnel evaluation. This category includes personal objectives, the linkage between performance and compensation, and evaluation reports. We grouped the firms in the first four years according to MCS intensity in the respective field. The employee curves for both groups categorized in year four are indicated in Figure 9.8.

The results show that high system intensity in the field of human resource evaluation in the first three years is not necessarily associated with high growth. Instead, in several years the performance of firms with low MCS intensity is even higher than the performance of companies adopting human resource evaluation MCS in years one, two, and three. Focusing on the grouping in year four shows that the picture changes. Now the growth rates of firms with high MCS intensity are higher.

A reason for the late need for human resource evaluation capabilities could also be found in the very nature of technology-based new ventures. MCS in this field comprise the linkage between performance and compensation or performance objectives. However, in the early days, the management team is compensated with shares or stock options rather than with bonus

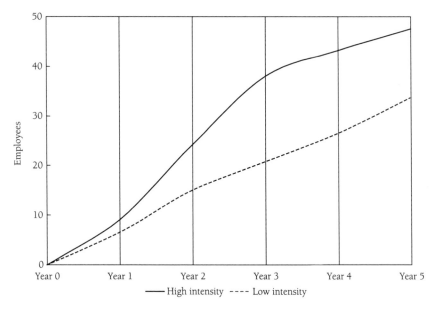

FIGURE 9.8. Human Resource Evaluation MCS Intensity in Year Four and Firm Growth

payments. This is mainly due to cash limitations. In addition, due to the high level of uncertainty, concrete objectives are difficult to define since priorities are constantly shifting. Thus, written performance objectives might not be met even though the employee performs exceptionally well. Only when uncertainties are reduced and more employees are hired do human resource evaluation capabilities become an essential element of a business organization.

Product Development Management Capability

The next group of systems focuses on MCS for product development. This group comprises MCS such as budgeting development projects, defining product development milestones, and evaluating progress, as well as quality and progress monitoring. We split the sampled start-up firms into two groups according to their MCS intensity in this field. Again, we performed the clustering in the first four years. The size over time of both groups clustered in year four is denoted in Figure 9.9.

In general, firms with high system intensity do not show a substantially higher growth rate than firms with a low intensity. Instead, in the first three years, new ventures with low MCS intensity in the field of product development perform better than firms, which adopt systems early. Only in year four is the employee curve of the high intensity cluster slightly above the group

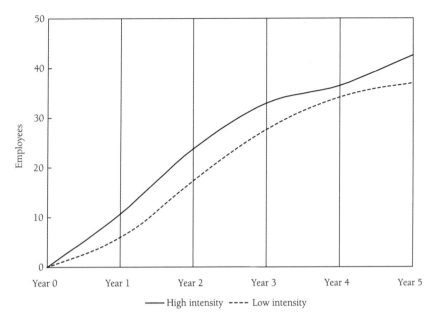

FIGURE 9.9. Product Development MCS Intensity in Year Four and Firm Growth

with less product development MCS in place. The mean size of both groups over time is shown in Figure 9.9.

The development of products is usually following the existence of a solid basic technology. As long as the entrepreneurial firm engages in experimentation and testing, capabilities to structure the product development process are generally not needed. On the contrary, the data seem to suggest that if the process of technological experimentation is formalized too early, creativity might be inhibited, which leads to reduced performance. The period of technology refinement and development usually takes a reasonable amount of time. Again, this is due to the high level of uncertainty. Only when the technological uncertainty is substantially reduced do product development capabilities become important.

Marketing and Sales Management Capability

The next group of MCS covers marketing and sales. This group includes elements such as individual target setting for salespeople, reports on open sales processes, or market research projects. We clustered the ventures into two groups according to the system intensity in the field of marketing and sales. Again we repeated the grouping for the first four years of existence. Figure 9.10 shows the employee development of both groups derived in year four.

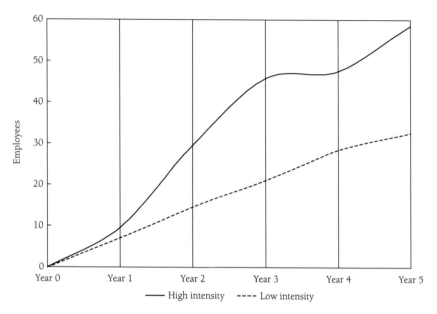

FIGURE 9.10. Marketing and Sales MCS Intensity in Year Four and Firm Growth

Grouping the firms in year one shows no substantial difference in terms of employee growth. However, when we categorize the companies in subsequent years, the outcome is different. In this case, new ventures with high MCS intensity show higher growth rates. This effect is especially strong for the clusters defined in year four.

Here, one might have assumed a different outcome. In general, products are developed before they can be marketed and sold. However, marketing and sales capabilities comprise market sensing and customer linking. Both skills are highly important for start-ups, even though they do not have a product yet. Without market sensing capabilities, an entrepreneurial firm might not even be able to develop a product that can successfully be sold. In addition, customer linking capabilities are necessary to win customers who engage in prototype testing and provide valuable feedback for the development of future products. Thus, the impact of marketing and sales capabilities on venture growth can be explained.

Partnership Management Capability

The last group of MCS considered in this study covers partnership management. This group comprises skills, which are related to business partners of technology-based new venture. The category includes partnership development planning and evaluation, corresponding policies, or partnership monitoring. We grouped the firms into high intensity and low intensity clusters in

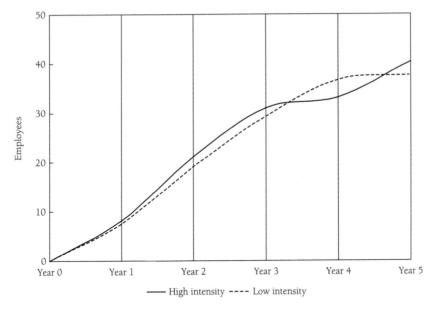

FIGURE 9.11. Partnership MCS Intensity in Year Four and Firm Growth

the first four years of existence. Figure 9.11 shows the number of employees over time for both groups categorized in year four.

Regarding the analysis carried out, companies that developed partnership management capabilities very early do not perform better than firms, which do not show these skills in the beginning. The results are very similar for categorization in years one, two, three, and four. In fact, in certain years firms with low system intensity show higher growth rates than new ventures with high MCS intensity in the field of partnership management.

Although entrepreneurial firms require partnerships, they do not need the skills to manage these alliances from the beginning. In the very early days, partners are mainly necessary to provide legitimization. In addition, partners can even be dangerous in the beginning since they might absorb knowledge that constitutes the competitive advantage of the entrepreneurial firm.

DISCUSSION

The results of this study comprise two aspects. First, the outcome shows that firms develop dynamic capabilities over time. The capabilities cover different management functions and do not evolve in parallel but at different points during the growth process of the start-up firm.

Second, the development of these capabilities is path dependent. Although new ventures could develop various capabilities in the very early days of their

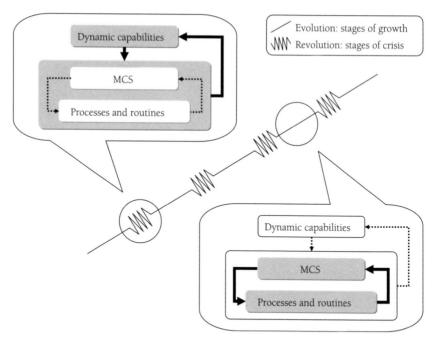

FIGURE 9.12.   Linking Evolutionary and Revolutionary Stages to Dynamic Capabilities

existence, they still might not obtain any benefits. On the contrary, the initial development of certain capabilities—like product development management and partner monitoring—may even compromise performance since it could impede creativity and block necessary resources. Only when the firm has reached a certain stage of development do new capabilities need to be added. At this time the available set of capabilities is no longer enough to implement growth, and the firm encounters a crisis that leads to the creation of new capabilities and further growth. Before this time the current set of capabilities is sufficient to implement incremental improvements, which are necessary to increase the efficiency of the organization.

This outcome of the study confirms Greiner's model, which asserts that evolutionary periods of incremental change are disrupted by discrete periods of crisis and revolutionary change. These crises occur when established practices become obsolete. Although the firm becomes more and more efficient within the existing routines, the effectiveness of the organization is reduced substantially. This is because the existing process and routines do not match the challenges the entrepreneurial firm encounters as it further increases its size. Thus, the organization requires additional capabilities to adapt the operating core to the new challenges. Crises usually facilitate higher-level learning in entrepreneurial firms (Cope 2005). In fact, shocks or jolts are necessary for "unlearning, new higher-level learning and re-adaptation to take place" (Fiol and Lyles 1985, p. 808). Consequently, the occurrence of a crisis fosters the development of dynamic capabilities. Having new capabilities in place, new ventures are able to change their resource base and implement new routines and processes. Hence, the organization is subject to second-order change.

The data suggest that the concept of capabilities is suitable to enlighten the growth processes and the transition between evolutionary periods and times of crises. Figure 9.12 combines capability learning with Greiner's idea of evolutionary and revolutionary change. In times of evolution, the emphasis is on single-order learning, the increase of efficiency, and the transformation of resources into economic rents. On the contrary, during periods of revolution the organization focuses on the development of additional capabilities.

CONCLUSION

This study contributes to the field of innovation management in that we developed a way to operationalize the concept of dynamic capabilities for fast-growing business organizations by establishing a link between the evolution of capabilities and the emergence of groups of management control systems. We derived a set of forty-nine management control systems representing eight distinctive dynamic capabilities. We were able to empirically observe the MCS in forty-four new ventures, relate the MCS to the development of dynamic capabilities, and further relate to the growth process of entrepreneurial firms. This approach can be deployed for future research not only in entrepreneurship but also in strategic management.

The study further contributes a longitudinal research design to observe the development of capabilities over time. All forty-four technology-based new ventures were studied over a period of ten years from their founding. In fact, while certain capabilities are beneficial in the very beginning, others are needed later in the evolution of entrepreneurial firms. Thus, we were able to distinguish successful from less successful growth paths along this very narrow differentiation of alternative paths. The set of eight managerial capabilities, which are based on forty-nine management control systems, can serve as (best) practices for venture creation. Time order of their implementation helps entrepreneurs as well as venture capitalists, incubators, and venture coaches in creating successful growth paths.

NOTE

1. One company in the sample was funded shortly after the study was finished.

REFERENCES

Alvarez, S. A. and Busenitz, L. W. (2001) The entrepreneurship of resource-based theory. *Journal of Management* 27: 755–775.
Amit, R. and Schoemaker, P. (1993) Strategic assets and organizational rent. *Strategic Management Journal* 14: 33–46.

Armstrong, J. S. (1982) The value of formal planning for strategic decisions: Review of empirical research. *Strategic Management Journal* 3: 197–211.

Bahrami, H. and Evans, S. (1989) Strategy making in high-technology firms: The empiricist mode. *California Management Review* 31: 107–128.

Barney, J. B. (1991) Firm resources and sustained competitive advantage. *Journal of Management* 17: 99–120.

Bierly, P. E. and Chakrabarti, A. K. (1996) Technological learning, strategic flexibility, and new product development in the pharmaceutical industry. *IEEE Transactions on Engineering Management* 43: 368–380.

Bourgeois, L. J. and Eisenhardt, K. M. (1988) Strategic decision processes in high velocity environments: Four cases in the microcomputer industry. *Management Science* 34: 816–835.

Brouwer, M. T. (2002) Weber, Schumpeter and Knight on entrepreneurship and economic development. *Journal of Evolutionary Economics* 12: 83–105.

Brush, C. G., Greene, P. G., and Hart, M. M. (2001) From initial idea to unique advantage: The entrepreneurial challenge of constructing a resource base. *Academy of Management Executive* 15: 64–80.

Bygrave, W. D. and Zacharakis, A. (2003) *The portable MBA in entrepreneurship.* Hoboken, N. J.: John Wiley and Sons.

Chandler, A. (1962) *Strategy and structure.* Cambridge, Mass.: MIT Press.

Churchill, N. C. and Lewis, V. L. (1983) The five stages of small business growth. *Harvard Business Review* 61: 30–39.

Cope, J. (2005) Toward a dynamic learning perspective of entrepreneurship. *Entrepreneurship Theory and Practice* (July): 373–97.

Davila, A. (2005) The emergence of management control systems in the human resource function of growing firms. *Accounting, Organizations and Society* 30 (3): 222–248.

Davila, A. and Foster, G. (2005) Start-up firms growth, management control systems adoption, and performance. Working Paper, Stanford University.

Day, G. S. (1994) Continuous learning about markets. *California Management Review* 36(4): 9–32.

Deeds, D. L., DeCarolis, D., and Coombs, J. (1999) Dynamic capabilities and new product development in high technology ventures: An empirical analysis of new biotechnology firms. *Journal of Business Venturing* 15: 211–229.

Dissel, M. (2003) Uncertainty and managerial decisions for new technology-based ventures. Dissertation, Faculty for Aeronautics, University BW Munich.

Dodge, H. R. and Robbins, J. E. (1992) An empirical investigation of the organizational life cycle model for small business development and survival. *Journal of Small Business Management* 30 (1): 27–37.

Eisenhardt, K. M. (1989) Making fast strategic decisions in high-velocity environments. *Academy of Management Journal* 32: 543–576.

Eisenhardt, K. M. and Martin, J. A. (2000) Dynamic capabilities: What are they? *Strategic Management Journal* 21: 1105–1121.

Eisenhardt, K. M. and Schoonhoven, C. B. (1996) Resource-based view of strategic alliance formation: Strategic and social effects in entrepreneurial firms. *Organization Science* 7: 136–150.

Evans, D. and Jovanovic, B. (1989) An estimated model of entrepreneurial choice under liquidity constraints. *Journal of Political Economics* 97: 808–827.

Fiol, C. M. and Lyles, M. A. (1985) Organizational learning. *Academy of Management Review* 10: 803–813.

Galbraith, J. R. (1982) Stages of growth. *Journal of Business Strategy* 3: 70–79.

Gompers, P. A. and Lerner, J. (1999) *The venture capital cycle.* Cambridge, Mass.: MIT Press.

Grant, R. M. (1996) Prospering in dynamically-competitive environments: Organizational capability as knowledge creation. *Organization Science* 7: 375–87.

Gray, B. and Ariss, S. S. (1985) Politics and strategic change across organizational life cycles. *Academy of Management Review* 10: 707–723.

Greiner, L. E. (1972) Evolution and revolution as organizations grow. *Harvard Business Review*: 37–46.

———. (1998) Evolution and revolution as organizations grow. *Harvard Business Review*: 55–64.

Grinyer, P., Al-Bazzaz, S., and Yasai-Ardekani, M. (1986) Towards a contingency theory of corporate planning: Findings in 48 U.K. companies. *Strategic Management Journal* 7: 3–28.

Gruber, M. (2004) Marketing in new ventures: Theory and empirical evidence. *Schmalenbach Business Review* 56: 164–99.

Gruber, M., Harhoff, D., and Tausend, C. (2003) Finanzielle Entwicklung junger Wachstumsunternehmen. In Achleitner, A.-K. and Bassen, A., eds. *Controlling für junge Unternehmen.* Stuttgart: Schaeffer-Poeschel.

Hanks, S. H., Watson, C. J., Jansen, E., and Chandler, G. N. (1993) Tightening the life-cycle construct: A taxonomic study of growth stage configurations in high-technology organizations. *Entrepreneurship Theory and Practice*: 5–29.

Hayes, R. H. and Pisano, G. P. (1994) Beyond world-class: The new manufacturing strategy. *Harvard Business Review* 72: 77–86.

Helfat, C. E. and Peteraf, M. A. (2003) The dynamic resource-based view: Capability lifecycles. *Strategic Management Journal* 24: 997-1010.

Henderson, R. and Cockburn, I. (1994) Measuring competence? Exploring firm effects in pharmaceutical research. *Strategic Management Journal* 15: 63–84.

Huber, G. P. (1991) Organizational learning: The contributing processes and the literatures. *Organization Science* 2.

Kamm, J. B., Shuman, J. C., Seeger, J. A., and Nurick, A. J. (1990) Entrepreneurial teams in new venture creation: A research agenda. *Entrepreneurship Theory and Practice.*

Kazanjian, R. K. (1984) The organizational evolution of technology-based new ventures: A stage of growth model. *Academy of Management Best Paper Proceedings.*

———. (1988) Relation of dominant problems to stages of growth in technology-based new ventures. *Academy of Management Journal* 31: 257–279.

Klofsten, M., Jonsson, M., and Simón, J. (1999) Supporting the pre-commercialization stages of technology-based firms: The effects of small-scale venture capital. *Venture Capital* 1: 83–93.

Kloot, L. (1997) Organizational learning and management control systems: Responding to environmental change. *Management Accounting Research* 8: 47–73.

Knight, F. H. (1921) *Risk, uncertainty and profit.* Chicago: University of Chicago Press.

Kogut, B. and Zander, U. (1992) Knowledge of the firm, combinative capabilities, and the replication of technology. *Organization Science* 3.

Kogut, B. and Zander, U. (1996) What firms do? Coordination, identity, and learning. *Organization Science* 7: 502–518.

Leonard-Barton, D. (1992) Core capabilities and core rigidities: a paradox in managing new product development. *Strategic Management Journal* 13: 111–25.

Levitt, B. and March, J. G. (1988) Organizational learning. *Annual Review of Sociology* 14: 319–340.

Lorange, P. and Scott-Morton, M. S. (1974) A framework for management control systems. *Sloan Management Review* 16: 41–56.

Lounsbury, M. and Glynn, M. A. (2001) Cultural entrepreneurship: Stories, legitimacy, and the acquisition of resources. *Strategic Management Journal* 22: 545–64.

MacMillan, I. C., Siegel, R., and Subbanarasimha, P. N. (1985) Criteria used by venture capitalists to evaluate new venture proposals. *Journal of Business Venturing* 1: 119–128.

Miller, D. and Garnsey, E. (2000) Entrepreneurs and technology diffusion: How diffusion research can benefit from a greater understanding of entrepreneurship. *Technology in Society* 22: 445–65.

Minshall, T. (1999) A resource-based view of alliances: The case of the handheld computer industry. *International Journal of Innovation Management* 3: 159–83.

———. (2003) Alliance business models for university start-up technology ventures: A resource-based perspective. *11th Annual High Tech Small Firms Conference*, Manchester.

Muzyka, D., Birley, S., and Leleux, B. (1996) Trade-offs in the investment decisions of European venture capitalists. *Journal of Business Venturing* 11: 273–88.

Nahapiet, J. and Ghoshal, S. (1998) Social capital, intellectual capital, and the organizational advantage. *Academy of Management Review* 23: 242–66.

Nelson, R. R. (1991) Why do firms differ, and how does it matter? *Strategic Management Journal* 12: 61–74.

Nelson, R. R. and Winter, S. G. (1982) *An evolutionary theory of economic change.* Cambridge, Mass.: Harvard University Press.

Nonaka, I. (1991) The knowledge-creating company. *Harvard Business Review* (November/December): 96–104.

———. (1994) A dynamic theory of organizational knowledge creation. *Organization Science* 5: 14–37.

Penrose, E. (1959) *The theory of the growth of the firm.* Oxford: Blackwell.

Peteraf, M. (1993) The cornerstones of competitive advantage: A resource-based view. *Strategic Management Journal* 14: 171–91.

Porter, M. E. (1980) *Competitive strategy: Techniques for analyzing industries and competitors.* New York: Free Press.

———. (1985) *Competitive advantage: Creating and sustaining superior performance.* New York: Free Press.

Prahalad, C. K. and Hamel, G. (1990) The core competence of the corporation. *Harvard Business Review* 68: 79–91.

Rangone, A. (1999) A resource-based approach to strategy analysis on small-medium sized enterprises. *Small Business Economics* 12: 233–48.

Ravasi, D. and Turati, C. (2005) Exploring entrepreneurial learning: A comparative study of technology development projects. *Journal of Business Venturing* 20: 137–64.

Roberts, E. (1991) *Entrepreneurs in high technology.* New York: Oxford University Press.

Rogers, E. M. (1962) *Diffusions of innovation.* New York: The Free Press.

Romanelli, E. (1989) Environments and strategies of organization start-up: Effects on early survival. *Administrative Science Quarterly* 34: 369–87.

Rothaermel, F. T. and Deeds, D. L. (2005) Alliance type, alliance experience and alliance management capability in high-technology ventures. *Journal of Business Venturing, forthcoming.*

Schoonhoven, C. B., Eisenhardt, K. M., and Lyman, K. (1990) Speeding products to market: Waiting time to first product introduction in new firms. *Administrative Science Quarterly* 35: 177–207.

Schuler, R. S. and MacMillan, I. C. (1984) Gaining competitive advantage through human resource management practices. *Human Resource Management* 23: 241–55.

Schumpeter, J. A. (1934) *The theory of economic development.* Cambridge, Mass.: Harvard University Press.

Scott, M. and Bruce, R. (1987) Five stages of growth in small business. *Long Range Planning* 20: 45–52.

Sexton, D. L., Upton, N. B., Wacholtz, L. E., and McDougall, P. P. (1997) Learning needs of growth-oriented entrepreneurs. *Journal of Business Venturing* 12: 1–8.

Shepherd, D. A., Douglas, E. J., and Shanley, M. (2000) New venture survival: Ignorance, external shocks, and risk reduction strategies. *Journal of Business Venturing* 15: 393–410.

Shrader, C. B., Mulford, C., and Blackburn, V. (1989) Strategic and operational planning, uncertainty, and performance in small firms. *Journal of Small Business Management* 27: 45–60.

Simons, R. (1994) How new top managers use control systems as levers of strategic renewal. *Strategic Management Journal* 15: 169–89.

Sinkula, J. M. (1994) Market information processing and organizational learning. *Journal of Marketing* 58: 35–45.

Smeltzer, L., Fann, G., and Nikolaisen, V. N. (1988) Environmental scanning practices in small business. *Journal of Small Business Management* 26.

Smith, K. G., Mitchell, T. R., and Summer, C. E. (1985) Top level management priorities in different stages of the organizational life cycle. *Academy of Management Journal* 28: 799–820.

Stinchcombe, A. L. (1965) Social structure and organizations. In March, J. G., ed. *Handbook of organizations.* Chicago: Rand McNally.

Stuart, T. and Sorenson, O. (2003) The geography of opportunity: Spatial heterogeneity in founding rates and the performance of biotechnology firms. *Research Policy* 32: 229–53.

Teece, D. J., Pisano, G., and Shuen, A. (1997) Dynamic capabilities and strategic management. *Strategic Management Journal* 18: 509–533.

Tushman, M. L. and Nadler, D. A. (1978) Information processing as an integrating concept in organizational design. *Academy of Management Review* 3: 613–24.

Tyebjee, T. T. and Bruno, A. V. (1984) A model of venture capitalist investment activity. *Management Science* 30: 1051–1066.

Ucbasaran, D., Lockett, A., Wright, M., and Westhead, P. (2003) Entrepreneurial founder teams: Factors associated with member entry and exit. *Entrepreneurship Theory and Practice.*

van Auken, H. E. and Carter, R. B. (1989) Acquisition of capital by small business. *Journal of Small Business Management.*

van de Ven, A. H. and Poole, M. S. (1995) Explaining development and change in organizations. *Academy of Management Review* 20: 510-40.

Verona, G. (1999) A resource-based view of product development. *Academy of Management Review* 24: 132–142.

Walsh, J. P. and Dewar, R. D. (1987) Formalization and the organizational life cycle. *Journal of Management Studies* 24: 215–232.

Wernerfelt, B. and Montgomery, C. A. (1988) Tobin's q and the importance of focus in firm performance. *American Economic Review* 78: 246–50.

Wright, M. and Robbie, K. (1998) Venture capital and private equity: A review and synthesis. *Journal of Business Finance & Accounting* 25: 521–70.

Zollo, M. and Winter, S. (1999) From organizational routines to dynamic capabilities. Working Paper, The Wharton School, University of Pennsylvania.

10

The Promise of Management Control Systems for Innovation and Strategic Change

TONY DAVILA

M anagement control systems (MCS) have traditionally been viewed as tools to reduce variety and implement standardization (Anthony 1965). They are associated with extrinsic motivation, command and control management styles, and hierarchical structures. Because their objective is to minimize deviations from pre-established objectives, they are designed to block change for the sake of efficiency. Learning comes from planning ahead of time, not from adapting to surprises. The functioning of a thermostat, in which a control mechanism intervenes when the temperature deviates from the preset standard, has been a frequent metaphor for this model (see Figure 10.1).

Not surprisingly, MCS are frequently perceived as stifling innovation. Therefore, their relevance to innovation—where uncertainty, experimentation, flexibility, intrinsic motivation, and freedom are paramount—appears to be limited. Innovation is to be managed through informal processes such as culture, communication patterns, or leadership. Uniformity and predictability—the hallmarks of traditional systems—are at odds with the need for the rich informational environment required for ideas to spark, grow, and create value. Coordination and control based on shared values substitute the "rules and procedures" of MCS (Walton 1985).

Over the last decade, increasing evidence has questioned the validity of these views. Intense use of MCS has been found in complex and uncertain settings (Chapman 1998). Budgets are key elements during episodes of strategic change "as a dialogue, learning and idea creation machine" (Abernethy

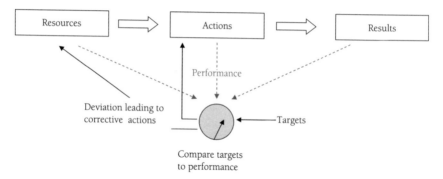

FIGURE 10.1. Feedback Mechanism Underlying Traditional Models of Management Control Systems

and Brownell 1997). The concept of enabling bureaucracy (Adler and Borys 1996) "enhances the users' capabilities and leverages their skills and intelligence" rather than with "a fool-proofing and deskilling rationale" typical of a traditional view. Companies exploit knowledge through flexible, user-friendly systems that facilitate the learning associated with innovation. Formal systems need not be coercive tools that suppress variation; rather, they support the learning that is derived from exploring this variation. Interactive systems (Simons 1995) have similar learning properties. They provide the information-based infrastructure to engage people in the communication required to address strategic uncertainties.

This chapter describes how MCS support different types of innovation and provides a framework to analyze their design. In developing this framework, it first examines how the concept of strategic process has evolved over time.

STRATEGIC PROCESS AND INNOVATION

The evolution of our knowledge of MCS is grounded on the progress that has been made in our understanding of the strategic process. Figure 10.2 summarizes this evolution and shows how innovation interacts with the strategic process along two dimensions. The first dimension is the origin of the innovation—whether it happens at the top management level or throughout the company. The second dimension is the type of innovation—whether it incrementally modifies the current strategy (incremental innovation) or radically redefines the future strategy (radical innovation).

Early concepts of strategy only considered the *deliberate strategy* (upper left quadrant), with formulation being followed by implementation (Andrews 1971). Top management formulated a strategy (deliberate strategy), which the company then implemented. In this view of strategy, MCS came in at the implementation stage to control deviations, much in the way in which the thermostat brings room temperature to its preset target.

Over time, researchers noticed that formulation and implementation happened at the same time and that deliberate strategy was not the full story. A new component was defined (Mintzberg 1978). Emergent strategy was formed as people throughout the company made day-to-day decisions. Strategy was shaped from the top but also from every person in the company as she adapted the deliberate strategy to her work. The existence of an emergent strategy led to the question of how top management could influence it (through, among other tools, MCS). The concepts of interactive and boundary systems (Simons 1995)—with the purpose of managing these "unexpected" decisions—captured this new role for MCS.

If day-to-day actions modify top management deliberate strategy, then why should top management go all the way to formulating it? The answer to this question led to the next step in the evolution of our understanding of the strategic process. Research suggested that top management does not formulate a deliberate strategy that is then randomly mixed with the emergent strategy. Rather, top management knows that the deliberate strategy is never implemented; instead of trying to force it, top management focuses on defining the guidelines that shape the emergent strategy (Burgelman 2002). The process of setting up these guidelines to induce certain strategic behavior is captured in the idea of *intended strategic actions* (lower left quadrant). These guidelines reflect top management's objectives rather than prescribe what the organization should do.

This idea was further refined through the observation that emergent strategy included two very different types of outcomes. Often emergent strategy evolved within the parameters of the current business model (incremental innovation), but sometimes it fully redefined what the company did (radical innovation).

Most of the time, strategy evolves through incremental innovations—as part of evolving objectives. These innovations are low risk, do not upset the existing strategy, organizational processes, or structures and systems, and build upon competencies already present in the organization or those that are relatively easy to develop or acquire. Even if incremental, these innovations are not necessarily cheap—incremental improvements in existing technologies and business models can be expensive propositions. Think about the cost of developing a new car—it is an expensive proposition, but in most instances it does not fundamentally change the way the company competes in the market. Moreover, if these innovations are well executed, they cumulate over periods of time into significant competitive advantage. For instance, consider the move of Japanese car companies from secondary to dominant players in the industry; relentless drive to do things better accounts for most of it.

In contrast, innovation that radically redefines the future strategy is high risk with high expected return; it significantly upsets organizations—shifting the power structure, redefining the relevance of core competencies, and requiring a redesign of the competitive positioning. These innovations are grounded on significantly different technologies and organizational capabilities, and depart from the current strategic trajectory of the firm.

Radical innovation is unpredictable, emerges throughout the organization from individuals or small groups with little if any awareness from top management, and is outside the current strategy; radical innovations are *autonomous strategic actions* (lower right quadrant). But lack of predictability does not mean that it should not be managed. Rather, top management has to put in place the soil for these innovations to happen and to be nurtured; MCS are among the tools to create this environment. Consider Intel's transition from a memory company to a microprocessor company. The shift into microprocessors did not start at the top of the organization; rather, by accepting and rejecting certain orders, developing the manufacturing technology, and designing the products, middle management shifted Intel's strategy toward microprocessors without much top management awareness. By the time top management decided to shift Intel's deliberate strategy, these products were already a substantial percentage of company sales. Other well-known examples include Post-it notes and NutraSweet. In these cases, radical innovations below the top management team were picked up by the company and transformed it.

But radical innovations are not limited to independent efforts across the company; top management itself can be the innovator. In the same way that top management shapes the current strategy, it can fully redefine the strategy of the company and become the source of radical innovations. The concept of *strategic innovation* (upper right quadrant) captures the idea of radical innovation happening at the top of the organization (Markides 2000). Consider Dell. While its success is associated with the ability to execute, its seed is a fundamentally new way of selling computers. Strategic innovation captures how strategy can be radically modified through the strategy formulation process that happens at the top of organizations. Strategic MCS (as they have been labelled) shape the information that top management has access to and become an important design variable (Lorange, Scott-Morton, and Goshal 1986).

A MODEL OF MANAGEMENT CONTROL SYSTEMS' DESIGN FOR INNOVATION

The previous paragraphs suggest that looking at MCS as hindering innovation is a narrow and obsolete view. Much to the contrary of this "conventional wisdom," these systems are very relevant to innovation. This section describes how various systems support the different types of innovation identified in Figure 10.2. Its purpose is to give a framework to analyze the design of the MCS of a company. Rather than looking at these systems as a whole, this framework helps identify whether the design of these systems is tailored to take advantage of innovation opportunities. Figure 10.3 pictures the framework.

Type of innovation

	Incremental innovation	Radical innovation
Top management formulation	Deliberate strategy	Strategic innovation
Day-to-day actions	Emergent strategy/ Intended strategic actions	Emergent strategy/ Autonomous strategic actions

Source of innovation

FIGURE 10.2. Types of Innovation Sources and Impact on Strategy

Type of innovation

	Incremental innovation	Radical innovation
Top management formulation	Deliver value	Build new competencies
Day-to-day actions	Refine current business model	Craft new strategies

Source of innovation

FIGURE 10.3. Four Roles of Management Control Systems for Innovation

Delivering Value: Management Control Systems for Implementing Strategy

The traditional role of MCS as tools to implement strategy is critical to delivering value. The quality of an idea is a necessary condition for success, but executing on the idea better and faster than competitors—that is, delivering value—separates winners from losers. The early success of Siebel Systems in the Customer Relationship Management market—Siebel reached 8,000 employees and $1 billion in sales faster than Oracle, Peoplesoft, and Microsoft—is based not as much on the customer relationship management (CRM) idea that was common to other start-up firms, but on its ability to be more efficient and faster than its competitors.

The relevance of these systems comes from their ability to execute efficiently and with speed. They simplify knowledge transfer—for instance, through standard operating procedures—and facilitate resources leverage—for instance, through delegation. Their strength, but also their weakness, is their effectiveness in translating strategic objectives into action plans, monitoring their execution, and identifying deviations for correction. In the process of enhancing efficiency, they potentially sacrifice the organization's ability to innovate.

Certain environments do not require much innovation, and a focus on delivering value does not give up much by forgoing flexibility. Standard procedures in power-generating plants integrate vast amount of knowledge in settings where small deviations may have devastating consequences. These procedures deliver the consistency and reliability to avoid costly mistakes.

MCS are the foundation of management by exception. Supervisors can delegate execution to subordinates, knowing that these systems will monitor and capture any deviation from expectations. These systems allow supervisors to reduce the attention they devote to activities managed by exception.

Another aspect where MCS are relevant to delivering value is accountability. Goals have a motivational purpose, and managers are held accountable to them. In contrast to standard operating procedures, here innovation is not blocked but is instead disregarded. Sales targets exemplify this argument; these targets are intended to motivate salespeople to deliver, thus ignoring any learning that may accrue for the individual salespeople.

Refining Value Creation: Management Control Systems for Intended Strategic Actions

In dynamic environments, new situations emerge that require innovative solutions outside the existing knowledge. MCS can be designed to capture and code these experiences to improve execution. These systems provide clear goals with freedom and resources for innovation and infrastructure to exchange information, search for new solutions, and gauge progress. The information in these systems evolves around the current strategy of the company and seldom leads to radical innovations. If designed to stimulate employees to explore new alternatives—through budgetary participation or what-if analyses—planning mechanisms such as strategic planning and budgeting advance the current business model and code this progress into expectations. Learning here is not as much anticipatory as experiential.

These systems refine existing organizational processes through innovation. For instance, companies invest in quality circles to gain competitive advantage through constant incremental innovations to current processes. Systems within product development can be designed to establish constant feedback mechanisms with potential customers (von Hippel 2001) to bring knowledge inside the company that stimulates innovation and translates it into a product. The nature of customer knowledge makes these innovations typically incremental.

Interactive systems—which top managers use to involve themselves regularly and personally in the decision activities of subordinates—stimulate discussion around the strategic uncertainties of the current business model (Simons 1995). Because they are defined by top management, they are more adequate for incremental innovation with the objective of making the strategy more robust to these uncertainties.

Product development manuals in two companies exemplify the two types of systems discussed. A first look suggested two companies with good processes in place, with stages and gates, clear procedures, and checklists to coordinate the support activities. However, the picture changed when talking with the managers of the process. In the first company, the manager saw her job as making sure that all the documents were in place, that every gate was properly documented, and that every step in the process was carefully followed. The objective was strict adherence to the procedures, which she saw as a blueprint to be copied. Deviations were exceptions that required corrective action. Her interpretation of the manual was a system to facilitate efficient product development, not as a system to capture and code new knowledge. Project managers saw her role as controlling of them. In contrast, the manager in the second company saw her role very differently. She sat down with project teams to tailor the process to the project's needs, to make sure that it provided value to the teams. The manager also reviewed each finished project with the project team to update the manual and make it more helpful the next time. Deviations were opportunities to bring about improvements to the current processes. The manual was alive, constantly evolving and incorporating learning. The product development manager saw MCS as tools to help execution and to capture learning opportunities that were lost in the former company.

Crafting New Strategies: Management Control Systems for Autonomous Strategic Actions

Radical innovations that lead to new strategies are more unpredictable than incremental innovation. They may happen anywhere in the organization, at any point in time. The process from ideation to value creation is much less structured, with periods when the path forward is unclear.

To generate radical innovations, companies need to (1) create the appropriate setting to generate ideas, (2) select among very different alternatives, and (3) grow new businesses. An important piece of this soil is culture (Tushman and O'Reilly 1997). However, the importance of culture does not imply that MCS are inadequate. Companies still need to think how to organize, motivate, and evaluate people; how to allocate resources; how to monitor and when to intervene; and how to capture learning in a setting much more uncertain and alien than the current business model.

Motivating people to explore, experiment, and question encourages new ideas. Strategic intent—the gap between current performance and corporate aspirations (Hamel et al. 1994), stretch goals (Dess, Picken, and Lyon 1998), and belief systems (Simons 1995) are potential approaches to motivate people to go beyond the current strategy. Strategic boundary systems (Simons 1995) focus these search efforts within certain parameters. Alternative approaches include internal processes, such as interest groups that bring together people with different training and experiences and external collaborations to explore alternative views. Access to resources—such as free time to "play" with new ideas—permits exploration of new ideas. Finally, the generation of ideas requires systems to exchange information so that promising ideas are identified and supported. The roles of "scouts" and "coaches" (Kanter 1989), or the concept of an "innovation hub" (Leifer et al. 2000) where ideas receive attention are examples of solutions through formal systems to managing autonomous ideas.

The selection of ideas to invest in also relies on MCS. But these systems are very different from the ones used for incremental innovation. Moreover, resources should be committed to each type of innovation prior to examining the investment opportunities (Christensen and Raynor 2003). Because of their higher level of technological, market, and organizational risks, and longer time horizons, radical innovations appear as less attractive than incremental innovations using criteria—usually financial criteria—applied to these latter types of innovations. Radical innovations require a selection process that relies to a larger extent upon the qualitative evaluation by different experts, generates commitment from various players in the company to make specific resources available, and has frequently been compared to venture capital investments. MCS are also needed to monitor and intervene in the project if required, to balance the tension between having access to resources and protecting the innovation from the current strategy that is designed to eliminate significant deviations, and to develop the complementary assets that the innovation requires.

Growing the business model also requires dedicated systems. The outcomes of a radical innovation are not limited to incorporating the innovation in the current organization—as for incremental innovation. Radical innovation can redefine the entire organization, become a separate business unit or a separate company as a spin off, sold as intellectual capital, or included in a joint venture (Chesbrough 2000). Moreover, the transition has to be managed carefully, especially if it becomes part of the existing organization, and MCS help structuring this integration through planning, incentives, and training.

Build New Competencies: Management Control Systems for Strategic Innovation

Top management is often the origin of radical innovations. Sometimes, these managers are the entrepreneurs who create the company; in other cases, they identify the need for a radical change and formulate the strategy to respond to this need. MCS support top management in (1) evaluating the need for radical changes, (2) formulating new strategies, and (3) building the competencies required.

MCS that support incremental innovation generally lead to refinements; but careful analysis can in some cases suggest radical changes. For instance, measurement systems such as balanced scorecards rely on maps of the current strategy (Kaplan and Norton 1996) and are mostly used as monitoring systems to track strategy implementation; however, they can also highlight opportunities for incremental improvements and for radical changes to respond to risks that threaten the current strategy. A similar analysis holds for any system used to monitor the current strategy, such as strategic planning systems, budgets, or profitability reports.

Stretch goals or demanding objectives are ways to create uneasiness in the organization regarding the status quo and a catalyst to search for radical ideas. Once rough ideas throughout the organization reach top management—through systems that move information along the organization from budgets to tools, such as second-generation suggestion systems (Robinson and Stern 1997)—can be powerful sources of radical innovations. Once the initial idea is formulated, experimentation with and exploration of the idea benefit from progress reports, analysis of external developments, and open questions to the future of the innovation.

Strategic innovation also benefits from MCS that systematically monitor the environment (Lorange, Scott-Morton, and Goshal 1986). Business opportunities emerge with changes in regulation, trends in customer needs, potential acquisitions, opening of new markets, or new technologies.

Ideas require further analysis involving local experiments. MCS also play a significant role in leveraging the learning required to formulate new strategies and building economic models such as scenario planning.

Learning for radical innovation contrasts with learning for incremental innovation. Incremental innovation relies on plans that work as reference points. But the knowledge that leads to these plans does not exist for radical innovation. Instead, MCS help to proactively manage the learning process. The planning involved does not outline specific milestones; rather, it lays out the motivation for developing new competencies, deploys the resources to build them, and puts together the measurement systems to define the new business model as learning evolves. MCS also structures a constant back-and-forth between vision and action through periodic meetings and deadlines to review progress. These deadlines pace the company and bring together different players to exchange information and crystallize knowledge. These meetings are comparable to board meetings in start-up firms. Board meetings pace the firm, force management to leave tactics and look at the strategy, and bring together people with different backgrounds to give a fresh new look at the company.

CONCLUSION

MCS are a key element in managing innovation and bringing it to strategy. This chapter frames this relationship along four lines: delivering value from the current strategy, refining the value creation process from the current strategy, crafting radical innovations throughout the company, and building new competencies for strategic innovations. Certain MCS are more attuned to the particular demands of each of these four roles, but they should not be seen as mutually exclusive categories. For example, the execution of a particular project—governed through systems to generate value—may raise some questions that lead to a radical idea. Similarly, systems to refine the current strategy may uncover a potential risk that leads to strategic innovation. Moreover, the importance of each type of system evolves as the strategy changes. Young strategies may require that organizations put more weight on systems for incremental innovation to accelerate the learning process. As strategies mature, the weight on these incremental learning mechanisms is expected to decay in favor of systems to implement strategy. Similarly, the emphasis on radical innovations varies with the success of the current strategy, with the location of relevant knowledge, and with the dynamism of the environment.

REFERENCES

Abernethy, M. A. and Brownell, P. (1997) Management control systems in research and development organizations: The role of accounting, behavior and personnel controls. *Accounting, Organizations and Society* 22: 233–49.

Adler, P. S. and Borys, B. (1996) Two types of bureaucracy: Enabling and coercive. *Administrative Science Quarterly* 41 (1): 61–89.

Andrews, K. (1971) *The concept of strategy*. Homewood, IL: Irwin.

Anthony, R. N. (1965) *The management control function*. Boston: Harvard Business School Press.

Burgelman, R. A. (2002) *Strategy is destiny: How strategy-making shapes a company's future*. New York: The Free Press.

Chapman, C. S. (1998) Accountants in organizational networks. *Accounting, Organizations and Society* 23 (8): 737–66.

Chesbrough, H. (2000) Designing corporate ventures in the shadow of private venture capital. *California Management Review* 42 (3): 31–49.

Christensen, C. M. and Raynor, M. E. (2003) *Innovator's solution: Creating and sustaining successful growth*. Boston: Harvard Business School Press.

Dess, G. G., Picken, J. C., and Lyon, D. W. (1998) Transformational leadership: Lessons from U.S. experience. *Long Range Planning* 31 (5): 722–32.

Hamel, G. and Prahalad, C. K. (1994). Competing for the future. *Harvard Business Review* 72(4): 122–29.

Kanter, R. M. (1989) *When giants learn to dance*. New York: Simon and Schuster.

Kaplan, R. S. and Norton, D. P. (1996). Using the balanced scorecard as a strategic management system. *Harvard Business Review* 74 (1): 75–86.

Leifer, R., McDermott, C. M., Colarelli-O'Connor, G., Peters, L. S., Rice, M., and Veryzer, R. W. (2000) *Radical innovation: How mature companies can outsmart upstarts*: Harvard Business School Press.

Lorange, P., Scott-Morton, M. F., and Goshal, S. (1986) *Strategic control*. St Paul, Minn.: West Publishing.

Markides, C. (2000) *All the right moves: A guide to crafting breakthrough strategy.* Boston: Harvard Business School Press.

Mintzberg, H. (1978) Patterns in strategy formation. *Management Science* 24: 934–48.

Robinson, A. G. and Stern, S. (1997) *Corporate creativity: How innovation and improvement happen.* San Francisco: Berrett-Koehler Publishers.

Simons, R. (1995) *Levers of control: How managers use innovative control systems to drive strategic renewal.* Boston: Harvard Business School Press.

Tushman, M. L. and O'Reilly III, C. A. (1997) *Winning through innovation: A practical guide to leading organizational change and renewal.* Boston: Harvard Business School Press.

von Hippel, E. (2001) Innovation by user communities: Learning from open-source software. *Sloan Management Review* 42 (4): 82–87.

Walton, R. E. (1985) Toward a strategy of eliciting employee commitment based on policies of mutuality. In Walton, R. E. and Lawrence, P. R., eds. *HRM trends and challenges.* Boston: Harvard Business School Press.

Index

About the Editors and Contributors

Tony Davila is a faculty member at IESE Business School, University of Navarra, and the Graduate School of Business at Stanford University, where he specializes in performance measurement and control systems for innovation management. He consults for large companies and Silicon Valley start-ups and has published in leading journals, including *Research Policy* and the *Harvard Business Review*. With Marc J. Epstein and Robert Shelton, he is co-author of *Making Innovation Work*.

Marc J. Epstein is Distinguished Research Professor of Management, Jones Graduate School of Management, Rice University, and was recently visiting professor and Hansjoerg Wyss Visiting Scholar in Social Enterprise at the Harvard Business School. A specialist in corporate strategy, governance, performance management, and corporate social responsibility, he is the author or co-author of over 100 academic and professional papers and more than a dozen books, including *Counting What Counts*, *Measuring Corporate Environmental Performance*, *Making Innovation Work* (with Tony Davila and Robert Shelton), and *Implementing E-Commerce Strategies* (Praeger, 2004), and co-editor and contributor to the multi-volume set *The Accountable Corporation* (Praeger, 2005). A senior consultant to leading corporations and governments for over twenty-five years, he currently serves as editor-in-chief of the journal *Advances in Management Accounting*.

Robert Shelton is principal at PRTM Management Consultants. He advises executives in a wide variety of industries and speaks on issues of innovation and business strategy to corporate, government, and university audiences around the world. He previously served as managing director at Navigant Consulting, vice president and managing director with Arthur D. Little, and managing director of the Technology Management Practice at SRI International, and his work has been cited in such publications as the *Wall Street Journal* and CNN Financial News and has been broadcast on NPR. With Marc J. Epstein and Tony Davila, he is co-author of *Making Innovation Work*.

W. Bernard Carlson is a professor at the University of Virginia, with appointments in the department of science, technology, and society (School of Engineering) and the history department (College of Arts and Sciences). Professor Carlson is an expert on the role of technology and innovation in American history, and his research focuses on how inventors, engineers, and managers used technology in the development of major firms between the Civil War and World War I. His publications include *Technology in World History* (7 volumes) and *Innovation as a Social Process: Elihu Thomson and the Rise of General Electric, 1870–1900*. With support from the Sloan Foundation, he is currently completing a biography of the inventor Nikola Tesla.

Steven C. Currall is professor of enterprise and the management of innovation and director of the Centre for Enterprise and the Management of Innovation at University College, London. He is also visiting professor of entrepreneurship and faculty co-director of the Institute of Technology at London Business School. Currall was formerly the William and Stephanie Sick Professor of Entrepreneurship at Rice University. His research focuses on commercialization of new science and engineering discoveries, public adoption of emerging technologies, trust and negotiation in inter-personal and inter-organizational relations, group decision-making processes within corporate boards of directors, and human resources/employment relations. He serves on the editorial boards of *Academy of Management Perspectives, Journal of Organizational Behavior*, and *Group and Organization Management*.

Bernhard R. Katzy started his professional career with an apprenticeship as a car mechanic. He is currently professor at the University BW Munich and Leiden University, and director of CeTIM—Center for Technology and Innovation Management. His research focus is on entrepreneurial management of fast-growing high-tech firms and the management of strategic change in the transition to the information age.

Kirsten Leute is a senior licensing associate in the Office of Technology Licensing (OTL) at Stanford University. She recently returned to OTL, where she started in 1996, after spending a year as a technology manager at the Deutsches Krebsforschungszentrum (German Cancer Research Center) in Heidelberg, Germany. She is a registered U.S. patent agent and is currently editor of the *AUTM Journal*.

Luis R. Mejia is a senior associate in the Office of Technology Licensing (OTL) at Stanford University, where he manages a portfolio of technologies ranging from electronics to marine biology. He has negotiated over 200 licenses in his 18 years at Stanford OTL and has evaluated over 600 inventions. Mr. Mejia is a co-founder of two Stanford spin-off companies, most recently Paraform, Inc. a 3-D software modeling company which was acquired by Metris International. Mr. Mejia is an advisor for Los Alamos National

Laboratory and the Monterey Bay Aquarium Research Institute and is a board member of the Stanford University OTL, LLC. Mr. Mejia managed the invention known as the PageRank Algorithm, which was the instrumental IP that led to the creation of Google.

Ramon O'Callaghan is professor, School of Economics and Business, Tilburg University, The Netherlands, and adjunct professor, IESE Business School, University of Navarra, Spain. His research specialties include information management and technology, innovation management, knowledge management, and strategic management. He is co-editor of *Transforming Enterprise: The Economic and Social Implications of Information Technology.*

Daniel Oyon is a professor at HEC Lausanne and a leading scholar in Europe on management accounting and control systems. Before joining HEC, he worked for several years as a consultant with Accenture and SAGE, serving mid-sized and large financial and industrial companies. His research interests focus on the importance of formal control systems on the execution and the emergence of a business strategy. In the past fifteen years, he has been in charge of several projects leading to the design and the implementation of modern management systems, including Activity-Based Costing and Balanced Scorecards.

Sara Jansen Perry is a doctoral candidate in industrial/organizational psychology at the University of Houston. Ms. Perry's research centers on technological innovation, collaboration, and strategic planning in research organizations. Before starting work on her Ph.D., Ms. Perry worked as a sales engineer for Sun Microsystems and SIS Technologies, two innovation-centric technology companies.

Florian Strehle joined 3i's venture capital team in 2005. Prior to this he was a management consultant with McKinsey & Company. At McKinsey Mr. Strehle was a member of the high-tech core group, focusing mainly on the semiconductor, telecom, and industrial technology sector. He specializes in IT and advanced technology investments across Europe, and participated in the Stanford Entrepreneurial Management Systems Project (SEMAS) on the European side.

Toby E. Stuart is the MBA Class of 1975 Visiting Professor at the Harvard Business School (2005–2006) and the Arthur J. Samberg Professor of Organizations & Strategy at Columbia Business School. He is also the academic director of Columbia's Eugene M. Lang Entrepreneurship Center. Professor Stuart's research has examined the formation, governance, and consequences of strategic alliances, the formulation of firm strategies in a number of industries, organizational design and new venture formation in established firms, venture capital networks, and the role of networks in the creation of new firms. In a recent project, he is examining the circumstances surrounding academic life scientists' technology commercialization initiatives, including starting and advising for-profit firms. He has served as an associate editor of the *American Journal of Sociology* and is a member of the editorial boards of *Administrative Science Quarterly, Management Science, Strategic Management Journal, Research Policy,* and *Industrial and Corporate Change.*

Kim Walesh is assistant director for economic and cultural development with the City of San Jose. She is also co-founder of Collaborative Economics, a Silicon Valley-based consultancy.

Kristi Yuthas is the Swigert Professor of Information Systems at Portland State University. Her research and consulting explore innovative ways in which companies can use information and management controls systems to improve organizational and social performance.